WILDERNESS, WATER, AND RUST

WILDERNESS, WATER, AND RUST

A Journey toward Great Lakes Resilience

Jane Elder

MICHIGAN STATE UNIVERSITY PRESS | *East Lansing*

Copyright © 2024 by Jane Elder

Part 1 epigraph comes from Jena Ball, "The Power of Storytelling," *Mother Earth News*, 2008.
Part 3 epigraph excerpt(s) from *The Immense Journey: An Imaginative Naturalist Explores the Mysteries of Man and Nature*, by Loren Eiseley, copyright © 1946, 1950, 1951, 1953, 1955, 1956, 1957 by Loren Eiseley. Used by permission of Random House, an imprint and division of Penguin Random House LLC. All rights reserved.
Part 5 epigraph scripture quotations are from The ESV® Bible (The Holy Bible, English Standard Version®), copyright © 2001 by Crossway, a publishing ministry of Good News Publishers. Used by permission. All rights reserved.

Michigan State University Press
East Lansing, Michigan 48823-5245

Library of Congress Cataloging-in-Publication Data
Names: Elder, Jane (Environmentalist), author.
Title: Wilderness, water, and rust : a journey toward Great Lakes resilience / Jane Elder.
Description: East Lansing : Michigan State University Press, [2024] |
Includes bibliographical references and index.
Identifiers: LCCN 2023022516 | ISBN 9781611864885 (paperback) |
ISBN 9781609177584 (pdf) | ISBN 9781628955217 (ePub)
Subjects: LCSH: Elder, Jane (Environmentalist) | Environmental protection—Great Lakes (North America) | Conservation of natural resources—Great Lakes (North America) | Great Lakes (North America)—Environmental conditions. | Environmentalists—United States—Biography.
Classification: LCC TD171.3.G72 E43 2023 | DDC 333.720977—dc23/eng/20230626
LC record available at https://lccn.loc.gov/2023022516

Cover design by Erin Kirk
Cover art is Adobe Stock / Brandon Schultz

Visit Michigan State University Press at *www.msupress.org*

For Colin and his generation

Contents

Preface

Back in 2005, my friend and long-standing Great Lakes colleague Dave Dempsey challenged me: "When are you going to write your Great Lakes book?" I laughed, but he had planted a seed. The Great Lakes region is home—a place shaped by five extraordinary lakes and the inland watersheds that feed them. The lakes—which hold one-fifth of the world's fresh water—shape our climate, our terrestrial and aquatic ecosystems, and the culture and economy stretching across eight states and two provinces. While the watershed (the Great Lakes Basin) spans 295,200 square miles, the region's forest composition and climatic influences extend well beyond that boundary into neighboring watersheds. Having worked in this region on environmental issues over much of my career, I had fodder for many stories. Still, old war stories and memories of special places aren't of much value to others unless they relate to a larger narrative.

A year later, Dave and I met for breakfast at a Great Lakes conference in Cleveland. As we watched the September sun warm the view of Lake Erie from the hotel windows, we talked about what we thought it would really take to save the lakes.

Dave said, "People have to tell their stories—about why they love the lakes, about why it matters. It's about what people value, how we choose to live. That's the level at which we need to connect with people if there's any hope of saving the lakes." So I set about considering what we value about life here, and how we choose to live in this place, in this time, as well as our complicated human relationship with the remarkable ecosystems that comprise the Great Lakes ecoregion.

So many people have powerful connections to these lands and waters. Like me, they care deeply about the freshwater heartland and safeguarding the region's vital ecological systems. This shared concern and commitment has driven extraordinary efforts to protect the Great Lakes over at least the last fifty years. I wanted to capture the stories of how our collective endeavors have shaped my journey, federal and global environmental policy, and the region's future at a critical time for the environment and our democracy.

Lakes and water are deep in the fabric of our sense of place in Great Lakes country. I love William Butler Yeats's poem "The Lake Isle of Innisfree," and especially this stanza:

> I will arise and go now, for always night and day
> I hear lake water lapping with low sounds by the shore;
> While I stand on the roadway, or on the pavements grey,
> I hear it in the deep heart's core.[1]

Yeats knew exactly what it is like to be deeply connected to a place. His lake was in County Sligo, Ireland. The lakes I love are in the heart of the North American continent: the lake of my childhood in southern Michigan, my favorite lakes in northern canoe country, and the urban lakes of Madison, Wisconsin. Then there are the giant Great Lakes in their vast watershed, spanning the borders of the United States and Canada as well the homelands of many sovereign Indigenous nations.

In such a vast region, our individual relationship with Great Lakes country comes from our own distinctive experiences. For some, it may just be that easily recognized blue graphic of the lakes on a map in "flyover" country and nothing else. For others, maybe it was standing on the shore and looking hard to see the other side and realizing you couldn't. It could be memories of wading into chilly water, watching a blazing sunset on a beach, or bobbing in a boat. Alternatively, it could be shoveling out from a lake-effect snow in Buffalo, Cleveland, or Houghton. Or, unfortunately, your lake experience might have been shaped by the sight and smell of mats of rotting algae, or cutting your foot on a zebra mussel shell on the beach.

Everyone's sense of place is different. But those who have that deep connection to the place that is the Great Lakes know it, even if it is hard to describe. That connection, based on love and respect for this natural wonder, is a source of power to sustain and nurture the lakes and the lives they support, and there's a lot to be done.

Love is powerful, but it isn't quite enough. We need to own up to the human impact of more than four hundred years of increasingly aggressive and callous use of lake country. Our dominant economic system has been built on exploiting the natural gifts of the region with little thought about the consequences for people and the natural environment. In the early twenty-first century, we find the broad Great Lakes ecosystem damaged by toxic industrial pollution and poorly regulated agricultural runoff, its food web disrupted by invasive species and degraded habitats. Pea-green algae often chokes western Lake Erie and Green Bay, and some still long to commodify or export the fresh water itself. A rapidly changing climate hangs over these threats like a dark specter, making all these existing problems worse.

Reiterating all the problems isn't enough to turn the tide, although it helps to understand how we currently find ourselves in the midst of vast ecological challenges. We need to take stock of where we are in order to move forward. And this much we know: the status quo will kill the Great Lakes ecosystem, so embracing change is the logical option.

We made considerable headway in the 1970s, 1980s, and early 1990s, but not enough. By the late 1980s, the policymaking process itself began to shift. Reaganomics, anti-government rhetoric, globalism, and the rise of corporate influence in politics all impeded progress. For the last several decades, we've been able to secure federal funding for Great Lakes cleanup, but we haven't had the political capacity to change the underlying threats. We keep paying to treat the symptoms but are unwilling to address the causes, or if we do, it is not at a scale that will solve the problems. This is not a path to sustainability or resilience.

It will take a sea change in "the way things are" to save the Great Lakes and all they do for humanity. Many long-accepted practices and systems aren't sufficient to protect the region's ecological health and that of its human communities—the old pattern is broken. If that's the case, then we also have an obligation to frame an achievable vision of how we can live in resilient harmony with this irreplaceable life-giving ecosystem. Rather than be daunted, let's explore what *could* happen to chart a new human journey in the region. There are signs of positive change already underway.

I reflect on how I learned to love and care for the natural world and lakes great and small. I look at the impetus for my activism and lessons learned along the way, including the challenge of wrestling with nature-scale issues with human institutions. After an intense fifteen years of work on Great Lakes land and water issues that began in the late 1970s, my career shifted to biodiversity conservation in the mid-1990s and then other areas a decade later. I served two terms on the International Joint Commission's (IJC) Water Quality Board between 2014 and 2020 while I was executive director for the Wisconsin Academy of Sciences, Arts & Letters. These experiences offered opportunities to take stock of what our earlier work had accomplished, but also to consider the sobering scale of what remains to be done.

Change *is* in the wind, and each of us will shape the direction it takes. I ask you to imagine, with me and others who love the lakes, what healthy ecosystems and a thriving society might look like in the freshwater heartland. Then, I ask each of us to dare to truly rise to the challenge of making it happen.

Acknowledgments

Many people made this book possible. First, I am deeply grateful for the encouragement and prodding of my friend and fellow Great Lakes writer Dave Dempsey. He urged me to write, helped me along the way over many drafts, and continued to believe that this was an important endeavor.

My husband, Bill Davis, lived with this effort from the beginning, as did our son Colin during his abbreviated lifetime. Without their support, this book would never have gone from concept to keyboard. Bill's professional knowledge in the field—especially in water policy—was a tremendous help.

I also want to thank those who read various drafts and offered helpful suggestions, especially Lee Botts, Tom Christie, Barb Sanford, Aaron Fai, John Jackson, Anne Woiwoide, Michelle Miller, Don Waller, Melanie Griffin, Dan Wise, Phil Barnett, Bruce Hamilton, Jim Ludwig, Eric Uram, Mark Van Putten, as well as various Environmental Protection Agency (EPA) staff and technical experts. Lifelong friends Kathy Menerey and Jane Moyer offered encouragement and insights into events from our lives. Bob Graves helped me recall events during the

Vietnam War, and other friends and family reconstructed timelines and experiences. I am deeply grateful that during the final months of his life, Al Combs urged me to go forward with the project after reviewing content about times shared at Pictured Rocks and in the Canadian Rockies.

The Mesa Refuge writer's retreat was a gift of immeasurable value—providing time and space to write the first draft, surrounded by writers—and I thank Cam Davis for opening that door for me. I am also grateful for the works of Kathleen Dean Moore and Terry Tempest Williams, whose compelling books helped inspire me to write about nature, love, and our human journey to protect this exquisite earth.

Prologue

Flint, Michigan, may seem like an odd place to begin a story about fresh water and the Great Lakes. But Flint is where my story began on a December day in 1954, in McLaren Hospital near the banks of a brown Flint River. I was a child of the late baby boom, born into a bustling community of postwar optimism fueled by the riches of the region's land and waters. Flint's story is also part of the larger story of boom-and-bust here—part of the story we need to acknowledge and learn from to live sustainably in a life-giving landscape.

Shortly after I was born, my parents moved from Flint to Pine Lake, just outside of Linden, Michigan, and my dad commuted to his job at AC Spark Plug in the city. On weekends we often drove to Flint to visit grandmothers, aunts, and uncles who lived nearby. But life on the lake seemed far away and apart from the city. Many in the extended family worked for General Motors, as did most of the people in the neighborhoods where our grandmas lived. The branches of my family came to Flint from farms and small towns for the good jobs in industry, as did tens of thousands of relocated families.

Their good jobs helped them buy tidy houses on tree-lined streets with sidewalks, where the ice cream man's bells would draw the children out like the Pied Piper on warm summer days. Grandma Elder's house had peonies and lilacs and a big cedar swing in the backyard upon which my brother and I would climb to listen to Grandma read Golden Book stories. Grandma Johnson had a white picket fence that framed a small garden with a big stand of raspberry bushes, just past the clothesline. Neither family lived lavishly; they weren't fancy people, and we weren't wealthy. But having come through the Depression and then World War II, for them life in Flint meant food on the table, steady paychecks, and the forty-hour workweek. Life was pretty good.

Downtown Flint was the big city to a country kid like me. When I was little, I was intrigued by trips downtown, especially if they included a stop in the big Woolworth's store, with its exotic parakeets and canaries for sale. The leading department store, Smith Bridgman's, was a special destination. We didn't shop there regularly—it was a bit tony for our budget. But it was the only place to acquire a Girl Scout uniform, and my mom liked their music department. So we had excuses to go.

A rare treat was lunch at the mezzanine restaurant. Just the word "mezzanine" seemed sophisticated and distinctive when I was a kid, and I remember my mother teaching me to cut my hamburger in half at that restaurant because it was more gracious for ladies to eat hamburgers that way. Smith Bridgman's was a symbol of the good life that Flint's working people could aspire to, and aspire we did. We were riding the economic surge in a company town—the same way steelworkers were riding the postwar boom in northwest Indiana, as were their counterparts in the chemical and paper industries, in addition to the other centers of manufacturing in the Great Lakes region.

Booms, by definition, are not sustainable. By the time I reached adulthood, the industrial facade was crumbling, leaving economic and social anguish in its wake. In the Flint area, in many ways, we were just playing out another scene in a pattern cast when the European explorers came to the region. They had been in search of a sea route to the "Orient," souls to save, lands to claim as their own, and, perhaps most significantly, riches to take away. Sustainability of the landscape and its people wasn't an objective.

Prior to European contact, the St. Lawrence and Great Lakes region was home to many tribes and nations within several language groups. The Iroquoian (Haudenosaunee) people lived along the St. Lawrence and the eastern basin, including Ontario. Within the Algonquian peoples, the Anishinaabe lived primarily in the northern lake regions, the Shawnee, Sauk, and others to the south, and the Menominee west of Lake Michigan. The western basin also included those in the Siouan language group, including the Ho-Chunk and Dakota. Territories changed with tribal disputes and sometimes brutal wars. Across this vast geography and distinctive cultures, the seasons shaped food availability and community life, and traveling to take advantage of seasonal food sources was part of many cultures. They hunted, fished, and gathered, and many cultivated food crops. The lakes and their tributaries provided extensive trade routes across the region. While Native peoples actively shaped their environments, for the most part they did so in ways that maintained the integrity of natural communities.

The long history of humanity in the Americas took a sharp turn toward unsustainable boom-and-bust cycles with European contact. In 1608, Samuel de Champlain established an outpost for France where the St. Lawrence River begins to merge with the gulf that embraces the Atlantic Ocean. This site, which would become Quebec City, became the launching point for further explorations via the waterways of the region, and later became a major center of trade. The French sent diplomats, traders, and missionaries to establish relationships with the people of the Great Lakes, including the Huron, the Ojibwe, and the Ho-Chunk. These emissaries also mapped the region and its resources. When Champlain first set eyes on the vastness of the salt-free waters of the Great Lakes in 1615, he called it La Mer Douce—the Sweet [water] Sea.

By 1673, Illinois guides introduced French explorers to the Mississippi River, and in their travels they completed a circuit linking two great continental watersheds. Through these journeys, the French discovered what the original residents already knew well: the lakes, their connecting channels, and their many tributaries were a very useful trade and transportation network that provided access to a vast array of natural resources. The French, and then the British, saw before them a land they considered untamed and theirs for the taking, setting the stage for the extraction-based economies that have influenced the lakes ever since.

The roots of these economies are colonialism and the exploitation of natural resources and people. During the boom part of this cycle, life seems filled with opportunity, and the beneficiaries of the boom are persuaded of the great wisdom of our species and its visible accomplishments toward what we label "progress." The bust part is ugly. If you take more from the land, the waters, and the people than you replenish, the system crashes and you rob the land and future generations of untold riches and resilience. If you focus only on production, not place, then when the product fails, so does the place.

The Europeans, and soon the new Americans and Canadians, built the first Great Lakes resource-extraction boom on the beaver trade. Between contact and the early 1800s, they trapped and traded millions of beavers and other fur-bearing animals, exhausting their numbers all the way from the Great Lakes to Alaska.

A family ancestor with the "Elder" surname first appears on this continent in a 1785 Pennsylvania birth record for a baby named Thomas. Who knows whether it was opportunity or necessity that impelled his progenitors—among the early Scots-Irish settlers—to leave home for a new land. They and others seeking land and opportunity helped swell the ranks of a new American people who would press deeper into the continent.

The great Shawnee chief Tecumseh allied his forces with the British during the War of 1812, in exchange for their promise of permanent homelands for the Indian people. In their battle over who would control the North American continent, neither the British nor the Americans held much respect for the sovereignty and autonomy of the Indigenous people. After Tecumseh died in battle, there was no champion to hold Britain accountable for its earlier promises to his confederacy when the United States and Britain negotiated the terms of settlement. Britain agreed to relinquish its claims to the Northwest Territory to the United States—territory between the Ohio and Mississippi River that included vast Indigenous homelands that Tecumseh had fought to protect. The interests of the First Nations of the Great Lakes region, and those in much of the rest of the continent, were thus disregarded and dispatched by the negotiators who wrote the Treaty of Ghent ending the War of 1812. As the signatories' ink dried on the treaty, the promise of a permanent homeland evaporated as well.

The door opened for settlement that flooded into the Great Lakes region. The fur trade soon collapsed, a victim of its own unsustainable practices and the changing whims of fashion. But it had created transportation and trading networks that supported new settlement. Towns grew from the old fur trading posts, and farmers carved out homesteads and fields in the landscape.

The century following the War of 1812 was a settlement boom and a resource bonanza. The lumbermen logged the great pine forests to the ground, and the sawmills turned them into building materials and ship masts. Commercial fishing on the Great Lakes expanded rapidly to feed the new population, reaching "peak fish" in the 1880s and 1890s for species such as whitefish, sturgeon, herring, and lake trout, which steadily declined thereafter.[1]

Like the Algonquin people before them, the new settlers and explorers found iron and copper deposits in the Lake Superior region. The mining boom began in the mid-nineteenth century. In the early twentieth century, a second timber boom turned the oaks, maples, and other hardwood trees into furniture and barrels. Rail lines and ports expanded to ship the ore to smelters and mills.

The growing economic opportunities attracted new settlers and immigrants. The Great Lakes region had a highly developed transportation system, productive farms, a growing and increasingly educated population, and abundant water and natural resources. Access to vast iron ore deposits and the ability to transport the ore by ship heightened its appeal to industry. Coal for blast furnaces was also available from nearby Appalachia. These assets made the Great Lakes a natural center for steelmaking. The lakes' transportation capacity gave rise to lakeshore steel-production facilities in New York's Tonawanda–Buffalo–Lackawanna region; Cleveland and Lorain, Ohio; the "downriver" area near Detroit; as well as Hamilton, Ontario, and Sault Ste. Marie. The largest steelmaking operations rose on the south shore of Lake Michigan, with Gary as its center.

Access to plentiful steel, water, and the ease of shipping coal to supply energy were all good reasons for the early automobile entrepreneurs to set up shop in the Great Lakes region. It also helped that Henry Ford was born in Dearborn, Michigan, a short hop west of Detroit. His family business bloomed where it was planted in Dearborn and kept growing.

My origins as a child of the Great Lakes are linked to the auto industry—the source of the biggest boom in the region. From Pennsylvania, my branch of the Elder family followed westward settlement into Kentucky and then southern Indiana over multiple generations. Just before the 1920s began to "roar," my Elder-Delashmit grandparents left their Indiana farm and came to Flint for work in the auto plants.

Like me, General Motors was born in Flint. The city's historic district is known as "Carriage Town" because it was a bustling center of carriage making at the turn of the twentieth century. GM started there in 1908 as a holding company of William C. Durant, the leading Flint carriage builder who made the leap to producing cars by buying out Buick, Flint's fledgling homegrown brand. The company grew through a flurry of mergers and acquisitions into the 1920s. Alfred Sloan took over in 1923. If Ford was the father of mass production, Sloan was the father of targeted marketing and planned obsolescence. In the Roaring Twenties, cars were becoming fashion, not just function, and the annual model change fanned consumer interest in what was in and what was out.

As GM grew, it made cars or parts in at least a dozen cities, including Flint, all in a humming network that produced a completed car at the major assembly plants. Walter Chrysler had risen through the ranks of GM and moved on to establish the Chrysler Corporation in 1925. In 1928, Chrysler acquired Dodge, rounding out the package that established Ford, GM, and Chrysler as "the Big Three."

As the region became the industrial heartland of the continent, more workers migrated from Europe, the American South, and other rural areas for jobs in the growing manufacturing communities. The jobs were there, but too often the owners treated the workers like mere cogs in the big machine. The miserable working conditions in the mines, on the railroads, and in the steel mills had already planted the seeds for a growing labor movement. So did the stark disparities between wealth and poverty.

In the U.P.'s copper country, a fledgling union walked out in a nine-month strike against Calumet-Hecla in 1913, in one of the classic struggles for workers' rights. The union fought for a whopping three dollars a day for pay, safer working conditions, and an eight-hour workday. When the union ended the strike, there were few gains and the owners rejected collective bargaining. While the unions rebounded in copper country over the next two decades, many of the high-quality copper sources in the Keweenaw were already depleted or headed in that direction. The boom was ending, and so were the jobs.

In this same era, my maternal grandparents, Joseph and Hilma Johnson, both immigrants from the Swedish-speaking region of Finland, found their way to the Lake Superior region. They had come to a new land when the settlement era was rapidly transforming into the modern industrial era. Joseph found work on the Lake Superior and Ishpeming Railroad, and purchased a home for $900 for his growing family in Marquette.

One of the few remnants I have of Joseph's life is a little notebook from the Finland Steamship Company, which must have been issued to the immigrant passengers traveling to the United States or Canada. He used the notebook to record things like work assignments and what he owed the grocer and other merchants. One page includes notes about an accident and illness policy he took out through the Benefit Association of Railway Employees in April 1924. This policy-holder-owned insurance company was an early example of how a pool of workers could organize for mutual benefit. For those of us who enjoy health, life, and disability insurance or are comfortable enough to take these things for granted, it's worth remembering that laborers long ago were pioneers in helping create benefits like these for working people and their families.

Joseph and Hilma lost their only son at the end of the Spanish flu epidemic in 1920. A decade later, when Joseph succumbed to another flu epidemic in 1930, there was a small widow's pension those allied railroad workers made possible. It helped my grandmother raise four daughters alone through the Depression. Through organizing, workers were stitching together a social fabric that could provide basic dignities of human necessities for themselves and their families in the face of a crisis.

During the Depression, the rapidly growing auto industry had no shortage of workers, and a manager could tell a disgruntled employee that if he was unhappy, there were plenty of other eager workers willing to take his job. Some conditions were ghastly.

In the 1930s, my grandfather Paul Elder worked in Flint's Buick assembly line in what its workers called the "pit"—a trench of sorts beneath a track over which the finished cars would pass with the engines running for a final tune-up before they rolled off the assembly line. The heavy exhaust gases would sink into the pit, where the men worked all day in greasy, noisy conditions. My dad said his father often came home with terrible headaches and nausea, with no choice but to go back the next day and do it all over again—that is, until the sit-down strike at nearby GM plants in town.

At the very end of 1936, after a period of heightened tensions, the workers at Flint's Fisher 2 and Fisher 1 auto-body manufacturing plants sat down on the job and occupied the plants. Two months later, GM's president, Alfred Sloan, announced a substantial wage increase and recognized the union, which ended the strike. The United Auto Workers began its rise to a new level of power and authority in the industry, and the working conditions began to improve in the plants in Flint, including the one where my grandfather worked.

The Depression years were hard on the region's growing industries, but the industrial production capacity built in the first half of the twentieth century was vital to the United States' role in World War II. The war put the region back to work. The Great Lakes states were among the largest producers of the weapons of our warfare, from airplanes to weaponry, ordnance, and other supplies. The scale of production was unprecedented in human history, as the United States stepped up to become the supply depot to the allied forces, made possible, in considerable part, by Great Lakes resources and labor.

The war years rebuilt the economy and birthed technological leaps in fields from pharmaceuticals to metallurgy. The chemical industry mushroomed into the production of pesticides, plastics, and countless other goods. The lakes provided an ample supply of fresh water and access to cheap electricity, and the industry grew rapidly along the Niagara River, as well as in Midland, Michigan; Sarnia, Ontario; and East Chicago, Indiana.

The postwar needs of new homeowners, young families, and a growing workforce fueled the mainstay industries of the Great Lakes once again. In addition, the second-growth forests were producing young trees that weren't adequate for building materials, but paper mills could turn them into pulp and paper. Wisconsin, Michigan, and Ontario were emerging as world centers of papermaking. For the longest time, the industrial smokestacks pumped their unregulated waste gases and soot into the sky. Growing cities and industries flushed their liquid wastes (including sewage) into the rivers and lakes, but in the Great Lakes region, the cities on the lakes drink the water, and the fish we like to eat live in the water. While Chicago had engineered its waterways to carry wastes away from the lakes, hundreds of other cities didn't have that option, or didn't worry about it.

The lakes are so large that many believed you couldn't pollute them, given the dilution capacity of *six quadrillion* gallons of water. Dilution, however, only takes you so far. The cumulative load of decades of industrial waste, along with the sewage from growing cities and the chemical runoff and silt from the region's agriculture, was beginning to take its dramatic toll on the lakes.

The pollution, combined with overfishing and the arrival of invasive species such as the lamprey, was undermining the lakes' ecological health. When I came along in the mid-1950s, the downsides were beginning to manifest themselves in the Great Lakes environment. Lake trout began to disappear from the region's markets and dinner tables. Soon, Lake Erie was plagued with massive algae blooms and widespread oxygen depletion in its waters.

In the 1960s, pollution of all kinds was increasingly pervasive. I remember a trip as a young child where we drove past the massive Dow Chemical complex near Midland in the heat of a Michigan summer, long before cars had air conditioning. The stench was awful. Our eyes stung, and we had to choose between the stifling air behind the closed windows or the acrid and choking smells of whatever Dow was pumping into the sky.

I also remember walking home from school one day in the wake of a highway truck spraying the ditches where water often pooled alongside the road. We were told that they were spraying the miraculous mosquito-killing DDT. The misty spray had an intriguing smell, and we kids thought the whole operation was fascinating.

People were increasingly alarmed by changes in the Great Lakes. Rachel Carson published *Silent Spring* in 1962, identifying DDT as the primary culprit in massive bird die-offs, and it wasn't only the birds getting people's attention. I remember my uncle Mac, who loved to fish, being frustrated when authorities closed Lake St. Clair to fishing due to the mercury pollution from Canada's chlor-alkali plants just upstream.

The International Joint Commission (IJC)—the U.S./Canada commission that oversees joint decisions about shared waters along their common border—began a series of consultations on Great Lakes concerns in 1964, because the effects of pollution and other stressors were increasingly visible. In 1965, someone caught the last living blue pike from Lake Erie, and to the best of our knowledge, the overfished species vanished from the earth forever.

In 1966, the United States took a stab at controlling water pollution by passing the Federal Water Quality Act, an update to the Federal Water Pollution Control Act and a precursor to the Clean Water Act that would modernize pollution control a few years later. In 1967 (and periodically into the 1970s), waves of dead alewives piled and rotted on Lake Michigan beaches. Then Lake Erie crashed—some called it dead—punctuated by Cleveland's flaming Cuyahoga River in 1969.

Oil slicks and floating debris had caught fire on other industrialized Great Lakes rivers, but *Time* magazine's photos of the river on fire caught the nation's attention. Finally there was a public outcry, and people demanded action.

The United States and Canada responded. In 1970, Congress and the Nixon administration established the Environmental Protection Agency (EPA), which assumed primary authority for grappling with water pollution. In 1972, the United States and Canada established the Great Lakes Water Quality Agreement, a binational pact focused on cleaning up nutrient pollution and sewage in the Great Lakes. There's a famous photo of President Nixon and Prime Minister Pierre Trudeau signing the first agreement. It was a historic attempt by two nations to acknowledge that water and natural systems do not respect political boundaries, and it was a remarkable leap forward in crafting diplomacy that responded to ecological as well as political needs.

Both nations hunkered down to the job at hand. The United States passed a new Clean Water Act in 1972, and Great Lakes states began to ban phosphate in detergents. Canada also committed to a major pollution-control program. The sewage construction business boomed with a rapid expansion of modern sewer systems and treatment plants, including the one that served the homes on the lake where my family lived. Over time, all Canadian provinces and twenty-eight U.S. states banned phosphate detergents. Lake Erie began to recover.

What we were also beginning to see was that an economy built on processes that poison public waters and foul the air its workers and neighbors breathe was not sustainable. A society that pretends the riches of the earth will forever be theirs for the taking with no consequences, is blind, foolish, and doomed. Corporate leadership that can divorce itself from place and people in the name of business profit is not leadership.

The 1973 oil crisis was a wake-up call to a nation that couldn't conceive of oil as a finite commodity or that wasn't in control of world supplies. Gas prices rose sharply, and the United States headed into a recession. By the mid-1970s, Flint was heading into tough times. The domestic auto industry had grown plump and self-satisfied, and the European and Japanese automakers were trying to carve out a piece of the U.S. market.

As the economy shifted, expectations about the wide-open opportunities in an auto town didn't match the realities. Disparities were increasing between those seeking the good jobs and the availability of those jobs. Social problems began to rise. Racial fear and white flight were emptying the hearts of Detroit, Flint, and other industrial cities. My grandmothers' little neighborhoods were beginning to change. One by one, long-term residents were moving out, and community stability was becoming more fragile. Drugs became the focal point of an alternative economy, entwined with a rise in crime.

In January 1974, a desperate young man broke into my grandmother's house while my aunt was at work, in what the police assumed was an attempted robbery. You must be desperate to think you can get much money out of an eighty-eight-year-old widow in a modest neighborhood. She must have resisted—she was a tough-minded woman—and the evidence indicated a struggle. He stabbed her repeatedly and left her to die at the foot of the basement stairs. Her tragic death left an enormous imprint on my mother, her sisters, and all of us in the wider family.

Among all the other devastating consequences, there was no going home to that place again. The little white house was no longer the orderly home with floors scrubbed so clean in that inimitable Scandinavian fashion that meant you truly could have eaten off them. It was no longer the house that smelled like coffee and cardamom bread on Sunday afternoons. Instead, it was a symbol of tragedy, a bellwether of a neighborhood unraveling and a social fabric fraying into tatters.

The oil embargo and gas shortages had only been the first wave of the perfect storm hitting the industry. The imports were steadily gaining a deep foothold among American consumers. They were smaller and more fuel efficient than the U.S. fleet and competitively priced. Moreover, they were gaining a reputation for better reliability.

Chrysler was in such trouble by 1979 that it turned to the nation for a $1.5 billion package of loan guarantees that became labeled "the Chrysler Bailout." The folk singer Tom Paxton wrote the witty but biting song "I'm Changing My Name to Chrysler" with the line "when they hand a million grand out, I'll be standing with my hand out." In fairness to Chrysler, the company did pay back the loan.

As the industry absorbed the blow, the ramifications reverberated throughout the region. Workers who had been with the plants since World War II suddenly found themselves facing layoffs, early retirement, and major plant closings. The steel industry, a major supplier to the automakers, got its own triple-squeeze: a waning supply of cheap Great Lakes ore; less demand from its big customer, the auto industry; and its own aging mills couldn't produce steel to the tolerances that the newer plants elsewhere in the world had introduced to the market.

The tide had turned on the steel and auto belt, and in the 1980s we had a new name: the Rust Belt. Sloan had not only planned the obsolescence of cars after their peak "model year"; his amazing corporation, which had helped define the American good life, was now in the "greed is good" decade, rendering whole communities obsolete.

The broader challenges across the entire auto industry were amplified in Flint throughout the 1980s. Michael Moore's documentary *Roger and Me* takes on General Motors' actions and their impact on Flint during this decade, when GM laid off forty thousand workers in the community. Upheaval and despair overwhelmed the city. While his film is a strong critique of Reaganomics and corporate greed, the seeds of Flint's decline had been planted long before the Reagan era, as they also were in other company towns throughout the region.

After Grandma Johnson's death, the aunt who had lived with her moved to an apartment near Flint's Genesee Valley Mall. In another part of the city, my uncle Chet was still living alone in my other grandmother's old house following an industrial accident that left him disabled. Staying in the family house enabled him to live with a fair amount of independence. Intruders had broken into the house once or twice, and the family was genuinely concerned, but he wanted to stay in his home. Then intruders broke in again. This time they took his meager cash supply, attacked him, tied him up with his own phone cord, and left him bound, injured, and alone on the floor. The vigilant neighbors—some of the last of the old neighborhood—noticed that they hadn't seen him out and about. When they went searching, they found him alive and rushed him to emergency care.

Independent living was no longer an option for him, and he reluctantly moved into a nursing home. The family agreed to sell the house. At the time, my husband and I were house shopping in Madison. We looked at a nice little two-story bungalow similar to that house. The Madison bungalow was on the market for about $75,000. My dad sold Grandma's house—with the big front porch and the stairway with the oak newel post, and the yard with lilacs, peonies, and maples—for $7,500. This is what economic collapse and fear will do to property values, not to mention what they do to people.

Smith Bridgman's, the anchor of the downtown shopping district, was an early casualty of Flint's decline, closing after bankruptcy in 1980 and torn down in 1984. Gone too is the Industrial Mutual Association (IMA) auditorium building, Flint's center of civic activity, where I remember family outings for homebuilding shows, Holiday on Ice, and the Harlem Globetrotters. The IMA's origins were in providing accident and life insurance for Flint's workers, but its mission grew much larger to provide a wide array of community services that built the quality of life in the greater Flint community.

Although GM continued to make profits during the challenging 1980s, they didn't plow those profits back into Flint or product innovations. There was no plan B for once-bustling communities like Flint, Michigan; Youngstown, Ohio; Gary, Indiana; and other towns that began to represent the unfortunate Rust Belt label.

A union job in the auto plants was no longer the career choice for many young adults coming into the labor force, because unlike their parents' generation, you couldn't count on it being there for a lifetime of pay and good benefits. The bubble of this American dream had burst. The good old days weren't coming back to Flint.

The decline of a once-vibrant industrial community must be hardest to watch for those who helped make it shine in the first place. In the summer of 2004, while visiting my dad and stepmom, we went to a band concert on the old firehouse lawn in the town of Fenton. The Shiawassee River runs along the bank, and on summer nights various bands played to people assembled in lawn chairs while a flag flew overhead. It's about as perfect a slice of Norman Rockwellian U.S.A. as you get can get.

A local Dixieland band was playing to a crowd dominated by retirees. At the break, an old coworker spotted my dad and came by to chat. He talked with fond remembrance of dances with Glenn Miller and other big bands at the IMA auditorium and what fun they were. There was a wistfulness in his tone. It is one thing to know that a time has passed; it's another to consider that a place has as well. Witnessing it all slip away in a few decades seems too much to have asked from the generation whose labor made the great years a reality.

The blows to the industry kept coming. In 2005, gas prices soared during the Iraq War, while the nation faced a trillion-dollar deficit—not a good combination for reviving an industry that had built its recent strengths on robust SUV sales. And so, a new round of closures, layoffs, and economic troubles rippled through company towns, with Flint again on the front lines. GM had already made significant cuts at AC Spark Plug operations over the years, but it absorbed a big hit in this round.

My dad and I talked about it when I was back for a visit in the spring of 2006. "In my day," he said, "there were fourteen thousand of us there. We turned out a million spark plugs a day. Today there are only 2,400 left in the labor force. What will happen to Flint?" He then went on to say, "You know, I don't even think they're making spark plugs in this country anymore. All the machines have been shipped to China."

People can impale themselves on the grief of dying dreams and dying places, but there's little value in that practice. It offers no healing or hope. Maybe all that labor was like a massive version of a Tibetan sand picture—beautiful in the act of its creation, and temporary in its final execution. After the monks create a beautiful design, they mix all the colors and patterns together into "just sand" again, and then give it back to the earth as an act of healing. The monks' act is intentional and purposeful, unlike the neglect, decay, and ghostly remnants of cracked pavement, shuttered plants, and abandoned houses of the manufacturing towns.

So, I wrestle with how we honor the endeavor itself—the extraordinary human capacity to produce something on a grand scale from ingenuity, sweat, and persistence, and the pride in the union label, with the contradictions of legacy pollution, lost jobs, and shattered versions of the American dream.

If there is hope for places like Flint, and for the larger Great Lakes region and its environment, it is beyond the formulas that left it abandoned or damaged. The boom-and-bust formula has played out in the Great Lakes region. It has failed the ecosystem, the human communities like Flint, Gary, and Detroit, and dozens of mining towns, timber towns, and fishing towns.

When a company leaves behind the company town, people with grit and imagination will try their best to rebuild a functioning community because they value the place they call home, including its history. But in Flint and so many other industrial cities, the odds are stacked so high against these new pioneers because the old model is broken and the capital has been used up or gone elsewhere. Sometimes, grit just isn't enough.

Instead, can we imagine what wholeness looks like for people and communities and the natural world in this region, with its green landscapes, sparkling lakes and rivers, sweetcorn summers and winter wonderlands? Our hearts long to live near the banks of cold water, and the gentle rivers, lakes, and fields of this giving landscape. If we give back to the landscape, and not just take; if we build communities that that nurture, sustain, and lift each other, and not just the suits in the C-suites, maybe we can restore and nurture a sustainable good life in the region of the greatest lakes in the world.

To get there, it will take policy, political engagement, and a new vision for the region. Who better than those who love the freshwater heart of North America to create that shining new future?

PART ONE

Rising to the Call

Find something that matters deeply to you and pursue it. Question. Stand. Speak. Act. . . .

—Terry Tempest Williams

CHAPTER ONE

Pine Lake Almanac

People tend to think of the Great Lakes as the big five: Huron, Ontario, Michigan, Erie, and Superior. But the ecosystem is more than the big lakes and their connecting channels. The basin (the watershed) embraces tens of thousands of inland lakes, rivers, swamps, bogs, and vernal pools. I had the privilege of growing up on the shores of one of southern Michigan's little lakes.

Pine Lake covers about 160 acres in southern Genesee County, approximately twenty miles south of Flint, between the small towns of Linden and Fenton. Byram Lake is just a hill or two over to the northwest, and to the east, over another rise in the landscape, lies Silver Lake and its far-reaching arms. The retreating glaciers that once covered most of the Great Lakes region scattered dozens of other small lakes across this part of Michigan in the upper reaches of the Shiawassee River watershed. With headwaters near Holly, Michigan, the Shiawassee flows westward into the county that shares its name, then north to eventually join with the Tittabawassee River. Together, they form the Saginaw River, a major Great Lakes artery that flows into Saginaw Bay in western Lake Huron. Like every other lake in Michigan, Pine Lake is part of the circulatory system of the Great Lakes.

When I came into this watershed, my parents were living in a small upstairs apartment in Grandma Elder's house in Flint with my two-year-old brother. Mom had grown up within a few blocks of Lake Superior and its wide-open spaces, and I suspect the whole family was feeling cramped as they anticipated my arrival. My mother's version of the story was that she did not want to raise children in the city, and so they scouted land around Flint and settled on Pine Lake. After they bought the lot, the developer built the basement and the shell of the house, and my dad set out to build the rest of the house on weekends and evenings while working full-time at AC Spark Plug in Flint. When I was six months old, we moved into the not-yet-completed house, but it was summer and I suspect they couldn't wait to settle into a place of their own in the country, even under construction.

We were among the handful of first year-round residents on the lake. After glacial retreat, mastodons had grazed this landscape, and prior to European settlement the Sauk and Ojibwe peoples had lived or traded in this area. The Shiawassee watershed was good beaver-trapping country in the French era, then a major wheat-producing region in the early farming days of American westward expansion. Forests within the Saginaw River watershed were among the first in the region targeted for logging in the white pine timber boom. By the mid-1800s, a sawmill and gristmill put the small town of Linden on the map. Early in the next century, carriage building transformed Genesee County, and specifically Flint, into a manufacturing center, followed by the auto industry.

After leaving Indiana in the 1920s, Dad's family settled into one of the new Flint neighborhoods built for GM's swelling ranks of workers. In the 1940s, my mother's family moved south from Marquette to secure jobs in the auto industry. The oldest sister left first for a job with GM in the Detroit area. The others soon followed for jobs with GM in Flint, bringing my grandmother with them. My parents met and married a few years after the war.

The postwar housing shortage and affordable automobiles led to a new phenomenon: the commuting workforce. The mobility offered by personal cars opened the door for developers to convert fields and forests into neighborhoods and suburbs, changing the course of land use and transportation planning. Urban design embraced the car. The Interstate Highway System, funded by Congress in 1956, launched the largest highway expansion—forty thousand miles—in modern history.[1] All that mobility and expansion would come at a price, from fragmented landscapes to demolished city neighborhoods, to smog and the climate impacts of fossil fuel—none of which was likely on the public's radar when we settled on Pine Lake in mid-1955.

Large farmsteads had surrounded the lake for the better part of a century. The developer, for whom our road was named, had purchased the lands on the eastern shore of one of the lake's two bays. Much of the other bay's lakefront belonged to another owner, and the Boy Scouts owned a camp tucked on the southeastern bluff. Elsewhere, the Hnatow family still retained extensive acreage from their original farm that wrapped around the planned developments on the north half of the lake. We came to a landscape of fading orchards, old pastures, wood lots, and a blue gem of a lake tucked into a glacially carved bowl.

Our house rose on a westward-facing hillside among a scattering of summer cottages and isolated homes along the east shore road. It offered a view and plenty of room to explore but must have been isolating for a young mother with two children underfoot all day. She talked about spiders taking over the cool, damp basement, snakes in the tall grass, and efforts to keep us away from the poison ivy that grew vigorously on the hillside below our house. She also talked about living with my dad's perpetual construction.

One of my earliest memories is getting in trouble for disassembling a portion of the partially completed fireplace. Inspired by watching my brother play with wooden blocks, I set about unstacking the bricks my dad had carefully laid—quite a heft for a preschooler. Picture, if you will, my dad's face as he came into the emerging living room to find partially dried mortar and bricks strewn across the newly laid hardwood floor.

Mostly, though, my memories are about how great it was to be a kid on the lake. Mom's instincts about raising us in the country were right on the mark. Growing up with a view across water is a remarkable gift. Facing west, that view included daily sunsets, and we grew to know where the sun would set over landmarks across the lake in the various seasons. Looking west also provided previews of storms coming over the lake. In the summer, we would watch the clouds build and the sky darken, and we could often see the front of rain peppering the surface of the lake as it headed toward us. Every now and then we would try to gauge the timing and race up the steps from the "flats" by the lake to the covered porch before getting soaked. Sometimes we pulled it off.

Occasionally the storms were violent and dramatic. Southern Michigan is tornado country, and my folks were still living in Flint when the deadliest tornado in Michigan history swept through Beecher and northern Flint in June 1953. After weathering the storm in their basement, they emerged to find their Rankin Street home intact, but houses a few miles north in the Beecher neighborhood were scattered across the landscape. The storm killed 116 people and injured another 844. This left a permanent impression on my parents. As a result, our household took any hints of tornado weather very seriously.

I remember at least a dozen times when we headed to the back of the basement (the front was exposed to the open hillside) and my dad would stand sentinel, looking through the window in the basement door while we listened to the radio back by the washing machine. With all those trips down the stairs, I don't remember ever seeing a funnel cloud over Pine Lake, but we certainly saw some violent weather. To this day, I'm vigilant about heading down to my own basement when the sirens sound, but I miss the spectacle of watching a front charging across water.

The seasons shape life in the Great Lakes region, and the contrast adds variety and dimension to daily experience. The windows that looked over the lake filled with this ever-changing story and an annual cycle of life and activity. We knew the lake's moods and their implications. When we gazed at calm water and the sultry haze of a humid summer day, it beckoned a swim before the likely arrival of an afternoon thunderstorm. A crisp and clear October sky over skipping waves was a sign to finish fall chores before a hard freeze took hold with the next front, and the gunmetal gray clouds of a winter sky told us to prepare for a big snow.

Looking westward across the water, we saw a small clearing from a former boat livery, and tucked behind the trees, the silo and rooftops of the Hnatow farm buildings beyond the long, narrow island that divided the lake into its two bays. Marshes grew along much of the western shore, and behind them a row of trees hid a narrow, seldom-used dirt road. It was, from our perspective, a wild shoreline.

In spring, remnant apple trees bloomed among the tender greens of the budding hardwoods across the lake. Red-winged blackbirds would return to the marshes, along with a great blue heron. Every year, a flock of Canada geese would return to the lake soon after the ice was out. Our hillside, covered with scrubby juniper trees of assorted sizes, was the beloved habitat of a large flock of blackbirds that also returned every year. Once a year, a wave of migrating cedar waxwings would stop to rest and refuel in the junipers on their way to some other home. The grass would green up, lilacs would bloom, and summer was just around the corner.

Summer is the time of year when Midwesterners pack in as much living as they can. My mother planted flowers around the house and vegetables down by the lake. She loved roses, nasturtiums, snapdragons, marigolds, and morning glories, and there were always petunias. In the vegetable garden, she grew tomatoes, green beans, leaf lettuce, a rhubarb plant, and sometimes carrots and squash. Currant bushes grew along the fence. As a child, I had no interest in the riches of the vegetable garden. Regardless, my mom assigned me the job of picking off tomato worms, and I dreaded it. I also didn't like the sticky feel of picking green beans, let alone the prospect of eating them after the life was boiled out of them, but the flowers were a different story.

I was always eager to go to the two greenhouses in Fenton where my mom would pick up seed packets and a flat or two of annuals right after Memorial Day when it was safe to plant. I loved the color and smell, and the heat and humidity in the greenhouses. I learned the simple things of gardening from my mom—reading the seed packets, measuring the depth of the hole, considering whether we were planting in sun or shade, and watering.

The gardens were only a tiny slice of the riches of summer. The lake called. Early in the season, Dad would wade into the chilly waters up to his waist and put in the dock, with neighbors or kids recruited to help. Once the dock was in, the boats went in. The big green boat was slightly larger than the average wooden rowboat. It could carry six people comfortably, sometimes more. Originally a wooden boat, over the years, others, and then my dad, layered it with fiberglass to keep it going. She didn't have much spit and polish, but she was a useful workhorse for summer on the lake. On a trip to Niagara Falls, my brother and I longed to go on the famous *Maid of the Mist* boat ride, but my frugal parents turned us down. Now, a child must do something with pent-up desire, and upon our return to Pine Lake, my brother named the green boat *Maid of the Muck*, and it stuck. Because she was big, broad, and stable, she was a fine boat for deep-water swimming.

Most of the nearshore areas of the lake were marshy and mucky—indications that we were on a "eutrophic" lake—one that was rich with nutrients, filled with warm water in the summer months, and destined to fill in with plants on its way to becoming land again. Limnology (the study of inland waters) calls the process by which lakes age "succession." Over time, lakes slowly fill in with vegetation and sediments, and change to marshes or bogs, and eventually land. Many mechanisms drive the aging process: encroaching vegetation, sedimentation, loss of water supply, or nutrients in the water. While succession is natural, nutrient infusions from human sources can push the accelerator. Runoff from fertilized farm fields, septic tank infiltration, and lawn fertilizers are all sources for the nutrients phosphorus and nitrogen.

In a warm-water lake, excessive nutrients fertilize weed growth and algal blooms. When the algae and weeds eventually die off, their decomposition uses up a lot of the dissolved oxygen in the water, which makes it hard for fish and other aquatic life to survive. This is what causes dead zones in places like Lake Erie, Green Bay, and the Gulf of Mexico. In small inland lakes, the effect is similar, plus the decayed leftovers of all those lake weeds and bloomed-out algae create more layers of muck.

For our human pleasures on the lake, we sought a pleasant swimming hole. Maintaining one took considerable effort—sometimes a bucket-brigade hauling out muck to repurpose as fertilizer for Mom's garden, sometimes using prop-wash from a boat motor mounted on the end of the dock to "blow out" the muck. The result was a small sandy-bottom swimming hole off the end of the dock. We also had a septic tank in those early years, quietly contributing nutrients to the muck-making process.

In the years after we moved in, many new neighbors who also commuted to Flint built houses on our shore. Others converted the summer cottages of my toddler years to year-round homes. One benefit of the growing neighborhood was that the active feet of kids playing in the swimming hole helped keep the muck at bay and the sandy bottom underfoot. But the muck was intent on closing in, and though we didn't suspect it at the time, so was the development.

The luxury swimming was in mid-lake waters, where the depth was twenty feet or more. My dad would load up the kids and friends in the *Maid*, motor out to the middle of the lake, and drop anchor. He'd bolt on the homemade ladder, climb on a seat, and dive into the water. I could rarely commit to the direct plunge, opting instead for first one step on the ladder, then the next, and finally easing into the water. A collection of inner tubes, life jackets, and air mattresses kept us afloat.

The boat was also our opportunity to explore the lake, and as we grew older, we could do so without a parent onboard. I loved poking around the marshy areas because they seemed so mysterious. Neighborhood legend had it that a giant snapping turtle lived in the weed beds by the northwest shore. We saw lots of turtles, but never the monster of the deep. By the shallow end of the island, we would row over a large patch of blooming water lilies. They swished and pulled underneath and alongside, and we imagined being in the *African Queen*, struggling to make it to Lake Victoria. The other bay had a mysterious bog ringed with tall trees and the earthy smells of acid soils and bog critters. You could always count on hearing frogs there. Behind the bog was an isolated stand of white pine that must have escaped the loggers, and down the shore, the summer camp.

My favorite destination was the rocks off the island. The island itself was a ridge of soils—a moraine left from a withdrawing glacier. At its eastern tip, the glacier left a group of boulders of assorted shapes. Most of these were submerged, but just barely. If you didn't know about them, they could ruin a boat or propeller, and a courtesy to new neighbors was to show them where the rocks were. Some good citizen improvised serviceable marker buoys out of empty bleach bottles tied to an anchor, and these let you know where the rocks were in years when they lurked just below the surface.

In high-water years, we would motor out to just beyond the rocks, cut the engine, and drift or row carefully up to the big ones and stare down into the water to see what was there. The biggest one—the boat killer—rose to a peak. We called it the "king rock." A court of smaller boulders surrounded the monarch. Rocks and water are just cool things for kids to explore in the fine pursuit of messing around.

The nearshore waters across from the rock bed included a gravelly bottom that bluegills and bass fish favored for spawning. Here, too, we could drift silently over the beds and look down to see the little plate-sized circles of cleared bottom surrounded by a ring of little stones. Sometimes, if the light was right, you could spy a fish patiently finning her tail, making a good place for bringing new life into the world.

Privately owned, the island itself was (mostly) forbidden territory. However, no one lived on the island, and hardly anyone ever set foot on it. Occasionally, we would make a beachhead, tie the boat to a tree, and explore. We discovered the ridge down the middle of the island, with a trail of sorts—delightful new territory for adventure. But summers were also the time for major house and yard projects, and we had to negotiate time for fun with things my parents deemed "productive," like painting the garage, washing windows, hanging the laundry, mowing the lawn, trimming hedges, and otherwise embracing the work ethic during the rare and precious days of a Midwestern summer.

As the *Maid* aged, we acquired a small aluminum boat, lighter and easier to row, and a good match for the little five-horse Johnson motor that I could start by myself with a good yank on the pull-cord. By the time the aluminum boat came into our lives—just before my teen years—nearly every buildable lot on the lake side of our road was filled. On the other bay, the same pattern was marching forward, and new houses filled two sides of the lake with year-round residents, with the island and the west shore as a buffer. The water still afforded wide-open spaces.

My girlfriend Kathy and I would take the boat out to the middle of the lake, anchor, and talk for hours. Sometimes we would sing in the safety of our isolation, and I will never forget Kathy throwing back her head and belting out the lyrics from the Temptations' "Ain't Too Proud to Beg"—"I know you want to leave me, but I *refuse* to let you go!"—after which we would dissolve into giggles, or ramp up a rendition of "Wild thing, you make my heart sing!" (Whatever happened to the Troggs?) We were just two small-town girls exploring the cool culture of Motown and rock 'n' roll in the sanctuary of a little boat on a little lake.

In our teen years, Dad brought home a small molded Styrofoam sailboat he had bartered in exchange for lawnmower repair. Although a bit beat up, it seemed functional. Mom hauled out her sewing machine, and we made a new sail for it and attempted to learn sailing with no formal instruction. We mostly succeeded, but we kept a sawed-off oar onboard for paddling home when we couldn't figure out how to beat against the wind.

Sailing a small boat is a different experience than puttering around in a more stable rowboat or power boat. This little craft held one passenger comfortably, and two who were good friends. You were close to the water and could drag a hand or foot along in it on a lazy day. It made you constantly aware of wind and wave direction and where you were in relation to your home dock. Sailing engages a fuller array of senses. It requires that you pay attention and reminds you with a flip and dunk if you slack off. It rewards you with the rush of wind and speed, with only the sounds of water humming over the hull. It gives you an improvisational dance on the water, new and different each time and very much of the moment.

On calmer days, when we tired of the focus that sailing demands, we could drop the little sail and just drift. Sometimes this was me and Kathy, and other times just me, alone. We would lie prone in the boat, floating without a destination. Just listening to the lap of water against the boat, feeling the sun on your skin, and hosting the occasional dragonfly or damselfly for a stopover—what could be better? For brief moments in summer, we were one with the wind, water, and sunlight. All was well.

Those idyllic summer moments were precious and rare. It is hard to be a teenager in any era. We were teens during a time of great social change. In the summer of 1967, the Detroit riots exposed the deep tensions in a community damaged by white flight and the early waves of a shrinking auto industry, and the 1968 protests at the Democratic Convention in Chicago heightened our awareness that the Vietnam War was raging.[2] The times were indeed changing on the campuses where we would soon be students, and changing in the wider world as well.

When Labor Day came around, school started, and the weather would soon change. Neighbors would lend a hand to haul boats out of the water and turn them over for the winter. Dad would stack the dock on the shore and drain the lawnmower's vital fluids and put it away for the winter. Mom listened for frost warnings and hastily gathered tomatoes if it looked like the growing season was ending. The view across the water would change to golds and rusty reds, and the yard would fill with the leaves from our ash trees and the neighbors' cottonwoods and willows that grew in the flats above the lake. The oaks that framed our view from the hilltop hung on to their leaves, browned and dried, well into the winter, and the wind would clatter through them and whine through the edges of the storm windows.

At some point, the geese would gather up and leave. The two "alpha geese" that the neighborhood kids called Honker and Beeper were easy to pick out. They would hiss and flap if we came too close, but they seemed like neighbors, and we got to watch them raise their children every year. Because we knew them, I always hoped they would leave before duck hunting season, but some years they lingered and we knew they were at risk. On a lake where the residents ringed two-thirds of the shore, those shotgun blasts always felt a little close for comfort.

Mom loved the crisp fall days, when the sky and the lake were a sharp blue and there was no trace of humidity in the air. She used to quote the first lines of a Helen Hunt Jackson poem:

> O suns and skies and clouds of June,
> And flowers of June together,
> Ye cannot rival for one hour October's bright blue weather.[3]

October's bright blue would soon change into November's brooding gray. There was no warmth in the afternoon sun or the short, dark days leading up to Thanksgiving. But the days after the holiday were the prelude to my early December birthday. My private birthday wish was always for snow, and I often got my wish. My mother dreaded winter because our poorly maintained dirt road turned and twisted around the hills above the lake, and the driving could be treacherous. She could worry about a predicted snowstorm like no one else. Winter meant isolation and risky driving to her, but to us kids it was a whole new season of fun.

Our hill was too steep for safe sledding, but the big one at the public park on nearby Byram Lake was wide, treeless, and long enough for an exciting run. My brother had a Flexible Flyer, the classic sled of our generation; I had a cheaper imitation. Sledding was fun, but ice on the lake opened an entirely new set of options.

I am the child of accomplished ice skaters. They had met on a community ice rink in Flint. Mom loved the Norwegian Olympic star Sonja Henie; Dad was a huge fan of the national and Olympic champion Dick Button. Their appreciation for the qualities of lake ice was sharpened by a skater's edge. We anticipated the first freeze, hoping for a good one. If it was windy and the lake was choppy, the ice would form in opaque white ridges near the shore, which was lousy for skating. If the air was still and the night very cold, one could wake up in the morning to a sheet of transparent, glassy ice, smooth and elegant, and ideal for skating. Dad's rule required four inches of ice before we could venture on the surface, which can take time in December. Often, a few nights of plummeting temperatures did the trick.

When the ice was exceptionally good, we would explore the lake through a new lens. You could lie down on your belly and stare into weed beds, spawning sites, or an abandoned beer can cold and still below. Where the water was deeper, there was just blackness below the ice—the dark, mysterious lake of winter. Sometimes, the lake would boom, and a compression crack would zig through the ice, relieving the pressure of expansion or contraction in the ice sheet. It was a huge basso profundo "ka-KUNG"—wild lake music on the bass clef.

I loved skating but was never really any good at it. When your parents are great at something, it can be intimidating for a kid to be less than great, especially in the awkward learning stages. Mom could cut figures (her favorite was a grapevine pattern) on the ice. Dad could spin and jump and did a fine spread eagle. I never even mastered the sliding stop, but I got around well enough with the skills I had—as did my brother, who had more interest in hockey than learning intricate moves. If there was no snow after a good freeze, we had the freedom to skate on the entire lake. These are rare gifts, so we took full advantage of the few days when it was possible. You had to watch out for cracks and ice fishing holes, but there was plenty of space to dream of Olympic glory, or just fly with the wind.

When conditions were less than ideal, which was most of the winter, we maintained a shoveled ice rink off our beach. Mom loved skating to music, nostalgic for Marquette's Palestra rink of her youth. Dad managed to rig a light pole and power outlet down by the shore, and we hauled our portable record player with its detachable speaker down to the lake, along with a few record albums, including *Nutcracker Suite*, *The Blue Danube*, and other selections lost to memory. It seems like a huge production to me now, but back then it just seemed like the logical solution to a challenge. I remember magical nights under the cone of light, skating in circles to waltzes as fat snowflakes fell from the black sky. When it was bitterly cold, the poor turntable would slow down and the music would groan and drag, but it was usually a sign that it was time to go inside anyway.

As the lake grew more populated, there were more ice fishers. In our bay, the serious anglers dragged out a handful of little shanties to promising spots over the deep water or the weed beds, and a whole community would spring up for the long winter months. Our family didn't ice fish, but as curious kids, we'd skate or walk out to the center of activity and check out how someone used an auger to make a perfectly round hole through the ice and then dipped out the re-forming ice with the little perforated scoop. We'd get to see stringers of fish, and watch "tip-ups" signal a strike on the line with their little flags.

People who don't ice fish muse about why other people do. Why sit in a little box on a frozen lake for hours hoping that a fish you cannot even see might happen by your dangling line and take your bait?

Here's my theory. There is an extraordinary quiet in the middle of a frozen lake, and a landscape (or icescape) like nothing else in your normal life. There's a slow, almost meditative quality to ice fishing: drilling the hole, setting the bait, and watching and waiting for a strike. You are surrounded by other people who *get* this, and you don't have to defend yourself. And if you are working a hard job, have a houseful of noisy children or a less than perfect marriage and just need to get away for a couple of hours in the cabin-fever months of winter, an ice shanty is an elegant solution. You aren't just loafing off; you are fishing for sustenance, so you are out there with a noble purpose. It's just you, the ice, and maybe a few fish, and like-minded buddies who will leave you alone with your fishing pole and maybe a six-pack, with a football game on the radio, and the rest of the world will pass you by. That's why people ice fish.

On winter nights, after we built the house addition and could look out over the lake through the bay window, I loved seeing a lantern or two blazing away in the middle of the lake while someone was storing up their peace and quiet late into the frozen night.

By late winter, which dragged well into March, we were as ready for spring as anyone. I sometimes joined Mom in the winter ritual of leafing through seed catalogs with all those overblown pictures that promised huge blooms and luscious fruit (and knowing that reality was a bit different). Seed catalogs on a dreary winter's day give you hope, and the seed marketers know this. That is how we ended up with cardinal bushes, several variants of rose of Sharon, the currant patch, dozens of rosebushes, and many other flowering shrubs in the yard, all of which made a respectable and appreciated effort to live up to their advertised splendor.

Eventually, the ice would take on a navy-blue cast, get mushy, and begin to melt and break up. One year, an "out like a lion" March wind drove the rotten ice onto the shore, pushing and piling long ice crystals like pure quartz into drifts on the shore. We stood by the edge and listened to the tinkling music of the dying ice of winter.

Soon the geese would be back and another spring and summer ahead. In later years, when my dad lived alone in the house on Pine Lake, we would use these signs to mark the passing of time. I would call home and he would say, "The geese are back," and I would know that something was right with the world.

My mother called it a child's paradise, and in many ways it was. But as Joni Mitchell sings, "Don't it always seem to go, that you don't know what you've got till it's gone." The very things that attracted my family to Pine Lake attracted others, and as the lake lots filled in with more homes and families, the changes in the lake accelerated. The humorist Erma Bombeck astutely observed, "the grass is always greener over the septic tank," and all those septic tanks, including ours, were fertilizing not just our lawns but the lake. Nutrient loadings from human development and activity pressed the "fast-forward" on eutrophication at Pine Lake and in thousands of other lakes across the region.

Our lake was already filling in with muck and reed beds along much of the shoreline in my early years. Succession was marching forward apace. At one point the neighbors in our bay banded together to pay someone to dredge a channel so they could still get their boats out to the lake's deeper waters. In other years, the lake association forked out money to chemically kill off some of the rapidly growing weed beds. Both followed a common human approach to environmental challenges—treat the symptoms, not the root cause.

After Congress enacted the Clean Water Act (CWA) in 1972, we entered a new era.[4] This new law created the framework to regulate pollutants that factories and municipalities discharged into U.S. waters, including federal grants to fund a massive wave of sewage treatment plant construction. Our neighborhood was eligible for sewers and federal funding. Although there were local costs, the plan went through. The construction crew tore up a lot of lawns and made a terrible mess of the road, but the sewers may have saved the lake—at least for a while.

After construction, the neighborhood's sewage (and that in nearby lake-neighborhoods) found its way through a series of pumping stations to a treatment plant just outside town. Most water treatment plants are set up for two levels of treatment: primary (screening or settling out the "lumps and chunks") and secondary (removing the remaining suspended or dissolved biological matter using microorganisms and "good bacteria" as well as various filters and aerating processes). Some plants also include a disinfection process to reduce bacteria and pathogen levels to protect downstream waters such as a swimming area. Different plants use different processes to get these jobs done. The goal is to capture the oxygen-robbing organic matter (that also makes the water cloudy) and release the cleaner water to a river or stream at the end of the process. The remaining semi-solids are called sludge, and in systems that separate the organic matter, the engineers call the stabilized material "biosolids," which are sometimes repurposed as fertilizer.

Some sophisticated treatment plants use a third level, called tertiary treatment. This extra step can capture and remove nutrients from wastewater or target other specific pollutants. Tertiary treatment is expensive, and most of the United States' 1970s-era plants don't provide this level of treatment—especially those operated by smaller municipalities. In the era of forever chemicals, pharmaceutical compounds that don't fully break down in human metabolism, microplastics, and whatever the virus du jour might be, it is useful to know that the CWA was never designed or funded to address these things at your local neighborhood secondary-level wastewater treatment facility. You can take some solace in another law, the Safe Drinking Water Act, which sets standards for what is acceptable in your drinking water, but in water, what goes around is likely to eventually come around somewhere else.

Treated wastewater from the Linden plant was (and still is) released into the Shiawassee River, the beginning of a long and winding journey, carrying waters from other communities, farm fields, and industries, to eventually flow into Saginaw Bay. Locally, the sewers reduced the nutrient load from septic systems on the lakes. They also reduced the risk of bacterial contamination in our well-water, a concern as more homes wedged close together in the pursuit of lake frontage.

The lake—and the world—was changing, and a new framework for environmental policy was in its nascent stages. The year 1972 was a policy turning point marked by the CWA and the Great Lakes Water Quality Agreement. For me, it marked high school graduation and heading off to college, eager for ways to be part of this new environmental wave of action.

When you grow up with the rhythm of the seasons in full view, when you have raced thunderstorms, traveled in a boat powered by your own strong arms, lolled under summer skies on welcoming water, you have been given great gifts. You have been given wonder, and freedom, and room to grow. As I stood on the threshold of adult life, I would carry the memories and lessons of a little lake and her seasons in my heart's deep core on the rest of my journey.

CHAPTER TWO

Seeds of Conviction

What makes someone an activist? While living near a lake forged my connection with nature, it takes more than proximity to compel a commitment to protect the environment, or the conviction that you must. When people ask me how I chose to get involved in environmental issues, there are simple answers—like growing up with a love for the natural world in a family with a strong sense of right and wrong and just being outraged at pollution and environmental destruction on principle. But I don't think that's the full explanation. I also think we rise in subtler ways to cues in our formative years or at points of life transition.

Whether through indirect messages in the background noise of the times and cultures in which we live, or bold, attention-grabbing events, these cumulative influences shape an internal watershed that forms the divide between witnessing, turning away, or making the choice to act. With the distance of many decades, I can see the strands that formed the warp and weft of my fiber in ways I wasn't aware of at the time.

First, let's pause to consider what has happened to the word "activist" in our culture. Somehow an activist—one who chooses to act on social issues—has taken on a negative connotation. In my active fundraising years, foundation program officers advised me to remove the word "activist" and even "advocate" from grant proposals because these words somehow connoted something edgy, affronting, or radical that raised red flags with cautious trustees and thus could jeopardize proposals.

The opposite of activist would be "passivist" (not "pacifist"). Is being passive on issues that affect our earth, our lives, and our children more desirable than being active?

I know the cautious program officers were advising against the "-ist" and "-ism" parts of the word. They warned me away from portraying the proposed work as being ideologically driven. One of the definitions of "ideology" is "the study of nature and the origin of ideas," and that doesn't seem so frightening. Ideology gets troubling when ideas and belief systems become rigid or completely divorced from context and fact, which is a reasonable concern that could lead someone to paint activism with a broad brush and react with aversion. Even so, it is narrow and unfair to lump active civic engagement with blind zealotry.

Over the last few decades, the term "environmental activist" has become associated with a stereotype of a shrill person screaming at the wind about yet another affront to the earth. The environmental movement unwittingly let that image get branded in the public's mind. Opposing forces sought to reinforce a perception that environmentalism was outside the mainstream. We lacked the savvy and resources to successfully push back. We lost sight of the power of other images—moms and dads working hard to protect their families from polluted air, water, and lead paint; people with a passion for the vulnerable creatures of this world, from hummingbirds to humpback whales; anglers who dream of a dance of wits with trout in a clean stream; workers on the assembly line who want safe, healthy working conditions and a good life for their families.

These too are environmentalists, along with a colorful tapestry of people taking everyday actions to make the world healthier, cleaner, and more vibrant. Isn't this activism? Whether one intentionally conserves energy, calls an elected official about an issue, votes thoughtfully, or teaches a child to think critically about issues, he or she is intentionally *active* in embracing solutions to social issues in varying degrees. I urge us to reclaim activism not as a lightning-rod label but as a healthy sign of a thriving democracy. Wasn't John Adams an activist calling for independence from an oppressive British government back in 1776? Weren't the abolitionists activists calling for the end of slavery, or the suffragists activists for fighting for a woman's right to vote? Let's face it: activism is more American than apple pie, given that the ancestral lineage of pie apples originates in the mountain forests of Kazakhstan. Activists are the reason there's a flag with stars and stripes on it in which to wrap issues in the first place. Someone's got to do the work of democracy—why not us?

So, back to those seeds in my activism. My parents believed in hard work, voting in every election, the value of reading and education, and community service. For my dad, that service included volunteering with the Boy Scouts, and later in life, the VFW, and for my mom, it was church, the Girl Scouts, and the Big Sisters. My mother was not a mere "Sunday" Christian. We were taught that everyone was a child of God, and nature—God's creation—was a wonder to appreciate. Putting the needs of others before your own was also central to her practice. I can hear her now, repeating, as she often did when attempting to guide our priorities: "God first, others second, self—last." These principles shaped my sense of where I fit into the world, and the obligation of service attached to our participation in that world.

In my childhood and adolescent years, the civil rights movement was in full bloom, and people taking action were part of the daily news. My family subscribed to *Life* magazine, and I remember seeing pictures of the 1963 March on Washington and grasping in some small way that something momentous was happening in our nation. Later that year, while my fourth-grade class was listening to story hour on the Friday before Thanksgiving, Lee Harvey Oswald's bullets struck and killed President Kennedy. The word "assassination" became part of our vocabulary. In addition to the shock, this sudden, violent end to a presidency unsettled our sense of stability. The world was changing in dramatic leaps and bounds, but this was the "normal" of the times.

When another assassin took the life of Martin Luther King Jr. on April 4, 1968, I was in eighth grade. Two months later, yet another killed Robert F. Kennedy. After King's death, I collected news stories and made a scrapbook. I found the remnants when sorting through old boxes years ago, and it felt affirming to rediscover that at age thirteen, I'd been sufficiently moved by his work to preserve a small record of King's impact on the world.

King's ability to see *the dream*, to articulate it with such passion and purpose that it spoke to the ages, also spoke to me about the power of his vision. What was most indelible to me was that he didn't speak of hope for hope's sake; it was absolutely *knowing* that the evil of oppression must not and would not triumph, but instead, our best humanity can and shall rise and change society for the better. This is conviction.

A few years ago, I think it would have been awkward if I had started talking about invoking our fundamental convictions of overcoming injustice, violence, and disrespect for the diversity of life during an environmental strategy meeting. People would have dropped their eyes to their shoes and waited for this moment to pass so we could get back to talking about target audiences, recruiting influentials, and all the pragmatic tools in the activist toolbox. But for the many who have responded to the calls for racial reckoning galvanized by the tragic deaths and disparities that reawakened America's attention in 2020, these necessary conversations are starting to happen.

In the rugged times—and we long-haulers have lived through plenty—the dream and conviction are sometimes all that is left to help you rise from the ashes and carry on. Dream the dream, articulate it in tangible ways that speak to our highest humanity, and carry conviction in your heart. These are the fundamentals I learned from the ripple effect of Martin Luther King Jr., augmented by my mother's value of compassion for our fellow humans, and my dad's sense of duty and responsibility.

Different dreams, driven by advances in technology—especially the space race—also influenced my generation. The networks televised every NASA launch, and in my elementary school, we gathered around a single television monitor in the "multipurpose" lunch-gymnasium-assembly room to watch the rockets blast into space. We grew up with the space program and CBS television news anchor Walter Cronkite carrying our imaginations into space. I kept another scrapbook from that era filled with *Flint Journal* clippings about the first moon landing.

We were living in times when we thought we were going to create a just and fair society and we could reach for the stars. Limitation, incremental progress, and roadblocks along the way weren't part of the public imagination. However, if the larger public had been paying attention, it would have seen that even with the passage of the Civil Rights Act in 1964, there were still major barriers that civil rights activists and marginalized communities were living with every day. The ability to dream beyond what used to be possible is part of the feedstock of advocacy, but a dream without the awareness of what stands in the way can be merely wishful thinking.

Despite the promise of the times, there were many bumps in the road to boundless horizons. One of them was the Vietnam War, which had been escalating throughout my middle school years and was certainly no mere "conflict." We watched the grim parade of casualties on television news along with the horrors of war. A month after King's assassination, as my classmates and I were beginning to anticipate next year's transition to the new high school, Max Nimphie was killed in Vietnam. I didn't know him personally, but his younger brother was in our class. To the best of my knowledge, Max was one of the first young men from our community killed in the war, less than six weeks after he began his tour of duty. The school held a community assembly, and they placed his folded flag in the high school trophy case next to the minor monuments for Linden's football heroics. Every other boy in high school who would soon face the draft would pass by that flag every day.

Families, including mine, would gather in front of TV sets to learn where brothers and sons would end up on the cruel lottery that matched birth dates with the likelihood of being drafted. The war was a presence that colored daily life and options for our future.

We had been raised with the values of World War II, where there was an assumed rightness to the cause, a clear sense of duty to serve for the sake of freedom itself. The murkiness of the mess in Vietnam offered no such clarity.

Any response or choice was laden with conflict. This forced many of us to wrestle with what we believed in, what we stood for, or at least to face that we were conflicted about what was right and wrong. That internal examination is important to another strand of activism: resolve. It is one thing to favor or oppose a set of positions; it is another to have examined your beliefs and the evidence and come to a firm conclusion about what your heart, mind, and gut tell you is right. We don't always get that clarity, but when we do, few forces in nature can alter it.

Despite all that was going on in the world, I was still dealing with the usual teenager challenges as a tall, skinny, good-grades, and not in the in-crowd teenager. I was in the role that was cast back in early elementary school—shy, studious, definitely not athletic—the nice girl in the back row of the class picture. During my sixteenth summer, I had the opportunity to spend a week at a church-affiliated camp in the Colorado Rockies and went for it. I fell in love with mountains, alpine meadows, and the exquisite blue sky, and had a crush on a nice guy named John, who I hope has had a wonderful life, wherever he ended up.

When you are a gawky teenager from a small town where you feel typecast, the opportunity to reinvent yourself with peers who know nothing of your labels and history is a powerfully liberating thing. In a short week, I learned I could climb a fourteen-thousand-foot mountain, enjoy a wild ride on a whitewater river, and meet and mix with people from around the country. I came home with new confidence and a stronger sense of self as I headed into my senior year in high school.

Small towns are wonderful in many ways but are also so constraining. Mr. Lowe, my high school guidance counselor, was genuinely concerned when he learned that I wanted to go to one of Michigan's two big state universities. At a time when feminism was spreading its wings, he, in his retirement year, was of the very old school. He wanted to steer me to community college and assumed that I would, at best, aspire to be a teacher. For many in his generation, teacher, nurse, airline stewardess, librarian, or secretary were still about the only career choices considered appropriate for "girls" to pursue—all fine professions, but a narrow list.

Perhaps he feared for me heading off to the unsettled campuses where demonstrations had risen to protest the war. In the fall of 1971, when we sat down to plot out my academic future, the May 4, 1970, Kent State University shootings were a fresh memory. At a campus peace rally, the Ohio National Guard had killed four unarmed students and wounded another nine, and reactions rippled throughout the nation. A reported four million students participated in walkouts throughout the United States. Later that same summer, the ill-fated Sterling Hall bombing killed a postdoctoral physics researcher and wounded four others at the University of Wisconsin. Unrest was a factor at colleges throughout the country. I suspect Mr. Lowe thought I was too green and fragile to handle a big school.

In truth, I sometimes wondered what I'd do in the face of a campus uprising, and I had some trepidation about a big campus with tens of thousands of students, but I knew I wanted to grab the brass ring. I finally persuaded him to sign whatever papers I needed to move forward. I learned at a reunion decades later that I wasn't the only young woman in my class who had been discouraged from pursuing their college ambitions.

For a variety of reasons, my family and I settled on Michigan State University, and we sent in the application. Eventually the fat envelope arrived, which was a good thing, because I didn't have a backup plan. Regardless of Mr. Lowe's attempts to steer me toward a "safe career," I was aiming for wider horizons. Blame it on *The Mary Tyler Moore Show*, and Mary's "spunky single woman succeeds in the working world" example, but I was interested in the world of television news and communications, as well as the environment and music.

My environmental interests were generic, rising in part from connections to Pine Lake and nature. In the wider world, we were also experiencing the broader rise of environmental consciousness in the country, from the ladies of Cleveland protesting the algae and pollution fouling Lake Erie, to the reports of choking smog in big cities around the country. The Earth Day "Love Your Mother" poster that hung on the wall in our biology classroom was a statement and a challenge at the same time.

When I considered pursuing environmental studies at MSU, the primary options were forestry or fish and wildlife management. There wasn't a crosscutting environmental studies curriculum at the time. Jobs in forestry were slim, and usually dealt with fieldwork and timber production, and fish and wildlife management studies were also largely about producing game species, which didn't seem like a fit for me. I concluded that there were plenty of technical people working in a small job market with forests and wildlife, but not a lot of people in the media telling the story about wider environmental issues. Although my interests could have taken me in many directions, my college form made you check a single box for "major." So I grudgingly picked just one, which is how I became a communications major.

Going from a town of 1,500 people to a campus with forty thousand students was like my camp experience in Colorado, except magnified exponentially. As a seventeen-year-old freshman, freed from the constraints of labels and small expectations, I could become whoever I wanted. And what a year for this to happen—campus was alive with the upcoming Nixon-McGovern election, public resistance to the war was growing deeper and more intense, and *Ms.* magazine had just started publishing regular issues a few months before my first semester. For a budding activist, coming of age in the fall of 1972 was a harmonic convergence of all the right elements.

I had opted to live in Asher House, an off-campus religious housing unit, which would have calmed all of Mr. Lowe's fears about my corruptibility at MSU. It offered a small community of about one hundred students in a dorm-like setting that was considerably quieter than the campus dorms.

The 1972 election season was an exciting time to be swept into the energy that pulses through a college town. While wondering if I had the smarts to succeed in my first semester with all these thousands of other bright students, I carved out some time to volunteer for the

McGovern campaign as well as the campaign of a young and energetic congressman from Flint who wanted to be a U.S. senator—Don Riegle.

The election took place just three weeks before my eighteenth birthday. Not being old enough to vote was so frustrating. On election night, a group of us gathered in front of the television in one of the Asher House common areas and watched McGovern go down in flames, which meant the war would continue, as would so much else that we feared.

In the numbing days that followed the 1972 election, the most committed activist in our circle challenged my despair. He told me that if I was serious about social change, I'd better understand that it was a long-term commitment, not just one election. He told me that I could sit around and wring my hands over the outcome of the presidential election or get back in the game and join an organization that was doing something important. He said, "If you care so much about the environment, join an environmental group." As the old saying goes, "Don't mourn, organize."

A few months later, I was visiting friends at the University of Michigan on a chilly winter weekend that coincided with the waning days of official U.S. involvement in the Vietnam War. The Paris Peace Accords had just been signed and were to take effect at 7 p.m. Eastern time on Saturday, January 27, 1973. At that hour, my friend Bob Graves and I were walking together past the big carillon tower that is a campus landmark, and it seemed that everyone was in a strange and somber mood. A single bell in the tower began to toll again and again and again. Someone on the sidewalk said, "The war is over. It's finally over." There was no rejoicing. Most of us just stopped and stood there, listening to the tolling bell, reflecting on the anguishing price this war had exacted from so many. After a time, we all quietly walked away.

In response to the weight of the times and my other friend's challenge to do something, I eventually worked up the courage to go to my first Sierra Club meeting on campus. I began to learn the practical steps of how people work together for a common cause. I went to slideshow lectures to learn about issues, gathered around long tables, and learned the enduring skills of stuffing envelopes and sorting by zip codes. From advice and experience, I learned the rudiments of how the state legislature and Congress worked.

In addition to hunkering down over the books and typewriter, I also had a part-time job at the campus photo lab, so I didn't have much free time to pour into environmental concerns or growing worries about a second Nixon term. But through the local Sierra Club meetings, I was learning more about participatory politics than in my political science classes. A side benefit was that the Sierra Club didn't only work on policy; it also encouraged fun in wild places, and I was increasingly drawn to the wild.

For so many of us troubled by a nation divided by the war, the wilds of nature offered something that seemed clean, sane, and "un-messed up" by the politics of our times. Camping stores were popping up in college towns, and John Denver was selling out concerts singing about a Rocky Mountain high. Nature and "natural" emerged as powerful new marketing labels of the day.

Having been smitten by the wilds of Colorado on my high school trip, my housemates and I organized a return to Colorado in the summer of 1973. We loaded up our buddy Tim Marshall's bright-yellow Chevy Nova and headed back to the same Colorado camps of my high school years for their collegiate week, with plans to also tour Grand Teton and Yellowstone National Parks.

We were about as straightlaced as any four college students could have been during the mid-1970s and were surprised at how people looked at us with a wary eye on our long drive to and from Colorado and Wyoming. This was a summer of increased tensions between the oil-producing nations in the Organization of the Petroleum Exporting Countries and the United States. Waiting in line for gas at stations with uncertain supplies was not unusual in the months leading up to the Arab oil boycott. Somewhere along Highway 80, we had to negotiate for a half tank of gasoline at a station that was rationing its short supply for its regular customers. A car full of college students just passing through wasn't good enough to qualify for a full tank of gas, and it made us fully aware of the serious gas shortages that summer and attitudes about students in more conservative regions of the country.

Still, we had a grand time in the mountains again at the camp's college week, which was punctuated by an odd and surprising visit from John Ehrlichman, who turned out to be a former camp trustee. One day, Ehrlichman spoke to the assembled college students in the camp's outdoor amphitheater, and expressed remorse for letting down young people through his role in the Watergate scandal.

In case your Watergate history is fuzzy, John Ehrlichman was Richard Nixon's domestic affairs adviser and considered one of the men closest to the president, along with H. R. Haldeman. Ehrlichman's Watergate role included establishing what became known as the "White House Plumbers" unit, a secret group of former CIA and FBI agents assigned to covertly stop the leaks from the White House inner circle—leaks, that is, about actions, legal or otherwise, to subvert agendas that were counter to those of the Nixon administration.

Ehrlichman's boss turned out to be a crook, and so was he in his time with Nixon. Still, his apology gave me hope that redemption is possible, even if repair is not. I thank him, too, for strengthening my resolve that democracy functions better when it isn't being "managed" by insiders who think they know what's best for the rest of us. I'll offer Ehrlichman some credit for his environmental work. He was one of the champions who persuaded Nixon to support the National Environmental Policy Act (NEPA), which gave our nation the tool of environmental impact statements, and the Environmental Protection Agency itself was established during the Nixon years.

After our encounter with a real-life co-conspirator, we set off for the Tetons and Yellowstone accompanied by another new pair of friends. It was a first-time experience for all of us, and one wow moment after another: sunrise on the face of the Grand Tetons, the thundering falls of the Yellowstone River, the wondrous geysers, and the thermal basins of Yellowstone—all in the warm fellowship of a band of friends united under western blue skies or gathered around a campfire. Politics, joyfully, was the last thing on our minds. It was hard to turn the car eastward again to return to our respective colleges and the realities of the times.

The following summer, thoroughly smitten with the wild outdoors from our western travels, my girlfriend Gaylynn and I attempted our first backpacking trip. With borrowed backpacks and new sleeping bags, we headed up to Michigan's Porcupine Mountains State Park. We of course found out that carrying everything on your back over long hauls and steep climbs is hard and challenging. We had done enough homework to know we needed to store our food away from the tent and suspend it from a tree branch higher than a standing bear could reach. The Porkies were known, even back then, to have some problem bears that went after human food. So, our first night—very tired—we made camp, made our dinner, cleaned up, and eventually succeeded in slinging a line over a branch and suspending our food bag.

As we settled into the tent and the darkness, we were startled by the screams of "Bear! Bear!" from the Boy Scouts camped at a group site nearby. Were we scared? Oh yes. We talked ourselves into some sense of security so we could get some sleep. "Yes, we washed the dishes and rinsed them far away from camp. Yes, we hung the food. Yes, we should be safe." Exhausted by the hike with heavy packs, we drifted off, with no signs of the bear.

Still, we made rookie mistakes. I will always be impressed with what raccoons and other smaller, curious critters can do to your gear if you leave it outside your tent. We learned the hard way what a raccoon can do to your one and only unattended water bottle. We awoke to discover that sharp teeth had punctured and drained most of our water bottle. Was it salty from our sweaty hands? Who knows what the attraction was. Wilderness travel invites (and sometimes requires) creative problem solving. Hiking without hydration in hot weather would have been foolish and miserable. We patched the bottle by melting a small piece of our nylon rope to fill the holes, which sealed them when it cooled, and repurposed a plastic bag to carry more water. Necessity can truly be the mother of invention.

The reward for the heavy packs and tired muscles is the quiet, and the beauty, the rare wildlife sighting, and so much room to breathe. In the Porkies, the trees are tall and majestic and the lakes clear and blue in the biggest wild place in Michigan. We covered a couple dozen miles in the backcountry on that trip and did it all by our scrappy little selves (aided by modern gear, of course). We encountered other backpackers on the trail who viewed us with raised eyebrows and said they couldn't believe two "girls" were out doing this on their own.

On our way home from our Porkies adventure, we spent the first night, August 8, 1974, in a state park campground in the western Upper Peninsula (U.P.). It was our first encounter with cars, campers, and electricity after our time in the backcountry. "Reentry" to roads, cars, and noise is a shock to the senses after a wilderness trip, but we welcomed a warm shower and easy access to plentiful drinking water. After we had finished cleaning up, we noticed people gathering around a camper that had a tiny little television glowing in an open door. As we walked up to the group to see what was going on, people were hushed and concentrating on the screen. We stood there, this odd assemblage of campers in the middle of a forest, and watched President Nixon resign. The following spring Saigon fell—the real end of U.S. involvement in the Vietnam war.

I graduated in 1976, the year of the American bicentennial. The nation seemed hungry for something to celebrate, and throughout the year various commemorations were a constant presence. We were saturated in red, white, and blue, but after Vietnam and Watergate, there was a fundamental shift in how we perceived our political leaders and our democratic institutions. Any post–World War II innocence about the infallible greatness of our nation—the stuff we were raised on—seemed tarnished.

As I left MSU, the nation was gearing up for grand celebrations on the Fourth of July, but my generation was embracing this celebration with few illusions. We had witnessed firsthand what happens when important decisions about war and peace and integrity in politics are isolated in the old guard. We had tasted the power of democratic dissent rising to challenge the status quo. We witnessed a crooked president finally forced out of office and a painful war finally end. Somehow the democracy did succeed. We believed that people could change the world, because we had, in some small but collective way, been part of that change.

I spent the bicentennial with someone I loved, camped on the shores of Lake Superior in the wild heart of Pictured Rocks National Lakeshore. We created a backcountry feast of pineapple-stuffed game hens that we roasted under the coals (the Park Service allowed campfires back then). As the sun set over this spectacular lake on the nation's two-hundredth birthday, we lit a single sparkler to celebrate all that was good about America, but especially the freedom to be in wild and beautiful places where the turmoil of the last decade seemed far away and small in the larger rhythms of sandstone, starlight, and clear waters lapping on the shore.

For many of my generation, participatory democracy would never again be "just voting." People talk about the 1960s as a decade of major social change, but the 1970s were a time when our democracy was strained, tested, and survived. In this decade, many of us in the "late boomer" generation would come to fully understand the old expression that democracy is not a spectator sport. The seeds of change planted in the upheaval in the 1960s were maturing into movements supported by staffed nonprofit organizations and finding agency in the mainstream political platforms.

Perhaps the wild country represented an unsullied ideal that deserved protection in its own right, or a place to breathe in strange and unsettled times, but my political convictions are linked with a yearning to be connected to nature. In a world where I was living through times shaped by napalm and Nixon on the one hand, and white pines and clear water on the other, it was an easy choice as to which is more authentic, enduring, and valuable for the sustenance of the human spirit and for affirming life itself.

I entered the adult working world with a rising conviction to apply whatever talents and skills I had in ways that would help protect, restore, and care for as much of the goodness of the earth as one small lifetime can achieve. It seems idealistic in retrospect, but it just seemed necessary at the time.

CHAPTER THREE

Into the Woods

Because we lived on a lake, there was little incentive for our family to travel to one for vacations. As such, my childhood contact with the five Great Lakes was minimal, and the northern forests, less so. The connections grew over the years along a path that wove in and out of a wide range of other experiences in the natural world and environmental politics. Still, a handful of visits in my early years left lasting impressions.

I was five years old when I first encountered Lake Superior. We had traveled to Marquette for my mom's class reunion, and stayed in a little resort nearby that had cabins shaped like tepees. Granted, this was not culturally in sync with the shores of Lake Superior, but my brother and I thought it was very cool at the time.

My mom seemed happy to be back on her old stomping grounds, and we walked to the nearby beach and stood on the wide, wondrous shores of this vast blue sea. You can't take a kid to a sandy shore and big lake and expect them to stay out of the water, even on a cool summer's day. I waded in up to my calves, and the biggest wave I had ever seen came in and drenched me in a shock of icy water. Welcome to the baptism of the north.

I shrieked, but it was partly in delight. As Katharine Hepburn's character in *The African Queen* says after the first run of rapids, "I never dreamed that any mere physical experience could be so stimulating!" I was intrigued, a little scared, and fascinated by this big, powerful lake. I still am.

Other first encounters included a trip to a Lake Huron state park where my biggest impression was the cobbled beach that hurt my bare feet trying to get to the water's edge. On a Niagara Falls trip, we watched the power of the Great Lakes thunder over the escarpment into the chasm below. This trip had the added excitement of traveling to another country and going through customs at the end of bridges named Blue Water, Peace, and Rainbow. Elsewhere in my memory is a trip up the western shore of Lake Michigan, with Mom in the driver's seat, traveling through the orchard and dune country. A highlight was stopping at a farm stand and picking up a quart or two of fresh cherries, which we demolished as the car hummed along the highway. I recall a day trip during Girl Scout camp to a vast Lake Michigan beach, and a family visit to a beachfront park on Lake St. Clair. But my deeper connection to the big lakes formed in later years as a natural outcome of the quest to explore and protect the remnant wild places along their shores and headwaters.

My trip to Pictured Rocks National Lakeshore had been my first shoreline backpacking experience along the Great Lakes, and I surely didn't want it to be my last. I was still, however, enthralled by mountains, and I signed up for a Sierra Club trip in the Canadian Rockies later that same month. I was trying to squeeze in as much exploratory time as I could that summer before I dove headfirst into a "real-world" job, and I'm so glad I did. The Canada trip was life-changing in a dozen ways. Few places on earth are as staggeringly beautiful as the glaciated peaks of the Canadian Rockies, and the backcountry was rugged and wild.

The itinerary described this "as a 'highlight' trip"—a play on the words "high," because we'd be in high altitudes, and "light," because pack horses would carry up to twenty pounds of our gear while we hiked. This is a luxurious way to travel by backcountry standards but no walk in the park. Our first task on the first day was to cross the swift but relatively shallow Miette River to pick up the trail. I remember taking off my boots, wading barefoot into water so cold it burned down to my bones, and thinking "This is crazy" before we were even a few hundred yards from the trailhead. In the days ahead, we had eighty more miles to go following a trail used on the first climbing expeditions to Mount Robson in the early 1900s. That initial stream crossing was a mere trifle compared to the challenges that lay ahead.

Over the course of our extended week, our little group formed a tight bond of interdependence. Careful planning, group cooperation, and a willingness to face fear and physical challenge on this ten-day trip were essential. Our most harrowing moment happened on day six. We needed to cross the flooded Moose River, and had to inch along a felled tree to reach the other side. One of the first of our group to attempt this slipped and fell into the river, clutching the safety rope our leaders had tied across the stream. The strong current pulled her off her feet, and the weight of her backpack pulled her under the water. Quick thinking, experience, and action by our leader, our packer, and his horse, made for a rapid and successful rescue. Not surprisingly, our hiking companion was shaken up, as were all of us. Those on the other side of the river rushed to get her in warm, dry clothes and offer deserved comfort and reassurance.

I hadn't crossed yet and had to use every ounce of spiritual training I had to overcome the fear of falling into the torrent, but I did it, and safely reached the other bank with a sense of triumph. That night, the hot soup and camaraderie in camp helped warm us and reassure us that we were safe and well and up for the other awaiting adventures.

We were richly rewarded for facing other stream-crossings over swollen rivers and slogging up muddy trails into high altitudes through the frequent rains that give these Rockies their green mantle. We drank from the crystalline springs flowing into icy streams in Helen Pass—a time before anyone needed to worry about giardia—and as promised by our packer, it was some of the best water I ever tasted. We feasted on fresh-caught trout for breakfast, glissaded down snowfields in the summer sun, and walked by the shining beauty of peaks with names like "Resplendent" and under the shadow of the imposing Mount Robson—Canada's highest mountain. We lolled in the splendor of alpine meadows in full summer bloom in Moose Pass and were swept away by the breathtaking vistas in the well-named Valley of a Thousand Falls.

This physically and mentally challenging trip stretched me beyond any previous sense of limits. I was also infused with a deep-seated joy from being fully immersed in the raw beauty of this place and the bonds of a small group of travelers who have shared a remarkable journey together. If this was what wilderness had to offer, I wanted more of it.

That "more" extended beyond the experience of adventure, wonder, and backcountry camaraderie. One evening around the campfire, our lead packer, Buster Duncan, talked about the need to protect these vast landscapes from modern development. He explained what he called "the next valley" assumption held by many of the early loggers, ranchers, and miners who came to the Rockies. They logged out many mountain valleys with abandon, with the justification that there was always "the next valley," but of course, there is a finite number of forested valleys fed by mineral-rich glacial streams, even in the Canadian Rockies, and all around the earth. I knew that the privilege of experiencing places like this came with the responsibility to care for them and protect them in some way, and this would shape my career for decades to come.

After that trip, I came home knowing I needed to get that first real job in the environmental movement. I had circulated petitions on campus during the spring of 1976 to place a referendum on Michigan's upcoming November ballot to require deposits on soda and beer bottles and cans. While reducing solid waste wasn't my passion, all environmental issues are connected, and this was a place to build skills and contribute to larger solutions. A concerted statewide effort had garnered enough signatures to put the question on the fall ballot, and the campaign was gearing up by the time we had come home from Canada. Michigan United Conservation Clubs (MUCC) stepped forward to manage the campaign under the imposing presence of Tom Washington, their executive director and the ultimate hook and bullet guy. I was thrilled to land a ten-week job on the campaign staff working to pass Proposal A, known as "the Bottle Bill." I thank Tom for giving me this career-opening opportunity, but I soon learned that he and I were from different planets.

While the unifying agenda was the Bottle Bill, conversations about other environmental issues were part of the office banter, and I soon discovered that my interests in wilderness issues were out of sync with Tom's and thus MUCC's. His notion of wild places was somewhere you could hunt bears. Given the fact that the Porcupine Mountains had been set aside many years ago as a wilderness state park with MUCC's support, you could, in his mind, put a check mark next to wilderness and be done with it.

Like Tom, some MUCC colleagues were openly dismissive of my Sierra Club leanings and my interest in protecting wilderness simply because it was wild and disappearing. It was my first encounter with what is purported to be a deep philosophical divide between conservation for the sake of protecting hunting and fishing resources, and conservation for the sake of broader environmental health, aesthetics, and intrinsic value. I thought the divide was artificial then and still think it is artificial now. It is more a cultural than practical difference—don't we all appreciate wildlife and want clean water and air?

At times, Tom treated the growing environmental movement as though it were some kind of threat to the long-standing conservation community grounded in the rod and gun clubs. I found it puzzling, as though in the uphill battle to protect the environment, there were too many laborers in the field. The environmental movement wasn't competition but rather an increasingly powerful ally.

Many new local and regional environmental groups were organizing; some were bringing professional staff onboard. There was talk of organizing a statewide coalition of groups including the West Michigan Environmental Action Council, the East Michigan Environmental Action Council, and others. Many of these groups, including the Sierra Club's state chapter, had joined MUCC in the spirit of the

Bottle Bill campaign but also saw a need to form alliances around their own interests.

Tom saw MUCC as *the* Michigan coalition on all things conservation and environmental and was hostile to the idea. I remember a staff meeting where he attempted to put an end to the idea of an alternative coalition by instructing staff to register a detailed list of likely coalition names with the state under MUCC's name so that no one else could incorporate under them. It was one of many times I observed Tom using his knowledge of process for power and leverage, and I was actively taking mental notes on how he knew and navigated the system. He was a master of the game, but I was also noticing how notions of fair play and partnership were fluid concepts subject to his interpretation. His tactics, while often brilliant and successful, also made me wary.

Tom was something of a living legend in Michigan conservation politics. As the new kid on the block, I didn't want to get on his wrong side. His behavior provided a dramatic incentive to learn how to cope or be consumed by subtle (and sometimes overt) harassment in the workplace. There were only a few women on the staff of a dozen or more at MUCC at the time. That included the receptionist and temporary campaign staff, like my colleague Cheryl and me. At MUCC, Tom made sure that you knew who the alpha male was, whether through openly expressing disdain for things you believed in (but he found ludicrous) or blowing a big puff of cigarette smoke in your face. He may have bestowed this behavior on the men in the office, but I never witnessed it. Learning how to navigate this bullying behavior came in handy in future work when I was dealing with the mostly male state legislature, Forest Service, Department of Natural Resources, and Congress.

Early in the campaign, some of my MUCC coworkers pasted a "God, Guns, and Guts Made America Free—at Any Price, Keep All Three!" bumper sticker on my car in the parking lot, next to my "Vote Yes on Proposal A" sticker. I wasn't amused. Still, while we didn't share a worldview, we shared a mutual goal to pass a statewide referendum that would reduce solid waste, save energy, and clean up litter. With our larger focus on that, things hummed along. I'm proud to say our passionate and hardworking campaign team, which included Tom and all the political muscle he brought to the fight, helped achieve a great victory for recycling and materials reuse in the 1976 election.

After the campaign ended, I joined the staff of the Public Interest Research Group in Michigan (PIRGIM)—one of the growing cadres of campus-oriented citizen-action groups inspired by consumer advocate Ralph Nader's work. PIRGIM was a better fit with my politics, and half of my new colleagues were smart and powerful women. Still, my heart was in environmental issues, which were only a small portion of the PIRGIM agenda at the time.

I remained an active Sierra Club volunteer, and met Brock Evans at a conference, then head of the Sierra Club's Washington, D.C., office and a passionate, inspirational champion of wilderness. He encouraged me to participate in the club's national volunteer training program and helped me secure a scholarship to offset my expenses so I could attend. My PIRGIM bosses let me go, figuring this was good training for the work I was doing there as well.

These were phenomenal programs. For a week, a group of about twenty volunteers from around the country dove into an intense Washington immersion experience. We toured the House and Senate office buildings and the chambers in the Capitol building. We met with members of Congress in hearing rooms. We trekked around the city and met with agency staff in the daunting halls of the huge buildings that house the Forest Service and Interior Department agencies such as the Park Service and Fish and Wildlife Service. We sought out and met with EPA officials in their unglamorous old headquarters beyond the highway interchange in Southeast Washington.

A highlight of the week was a meeting in the White House office building. This was during the Carter administration, and we were welcome in those days, but I still remember extensive security checks before we could enter the grounds and the realization that your nation was keeping a file on you.

An eye-opener was a briefing at the national headquarters of the Chamber of Commerce, where we had a glimpse of the tony offices and extensive resources of a lobby we often opposed on a range of issues. They were friendly hosts, in the way that I suspect trial lawyers on opposite sides of a case can be friendly in a professional way out in the hall and then tear each other apart in the courtroom. This was an important window into the world of the professionalized approach to lobbying, in which the substance of your cause can be divorced from the process.

When you are driven by your convictions and passions, it's hard to imagine applying your advocacy skills to things you don't believe in, but a lot of professional lobbyists make a fine living supporting someone else's cause because it can be viewed as simply a "job" that requires astute knowledge of the process and the players to succeed.

Lobbyists have a bad reputation as being sleazy powerbrokers. I used to have a button that said "Corporate Lobbyists Eat Their Young," which I still think carries some truth, but there's nothing inherently evil about the art and skill of lobbying. It depends on how you use those skills and to what end. Despite the caricatures, the average lobbyist, even on a corporate retainer, doesn't run around with wads of cash stuck in his or her back pockets. Successful lobbying, no matter the cause, is a blend of instinct and information. It requires knowing who talks to whom, who has what kinds of power and influence, and who cares about an issue or doesn't. It means understanding what a decision will cost in terms of public dollars and political capital, which can be anything from getting a good committee assignment to campaign support. It is also about understanding timing and process, from the Byzantine rules and procedures of committees, floor votes and federal budget calendars, to the closure dates on public comment processes and the deadlines for newspapers to go to print. Occasionally, it is about facts. But the facts only matter if you can navigate the rest of the territory, which is why lobbyists have value in the system.

A good lobbyist is a seasoned guide through the policymaking labyrinth, and they can place the precious substance of your cause in a position for a favorable outcome. Still, even the best package, skillfully delivered, requires a vote by the members of the legislative body. You can have the most expensive lobbyists on the Hill and still lose if more members of Congress vote against you than with you. Lobbying is part of the journey, but the destination is determined by those who eventually make the choice to vote yes or no, or the voters who choose to send those decision-makers to office in the first place.

Our final challenge of the week was to meet with our own members of Congress. The organizers wanted us to taste what it was like to walk those long halls in what I began to call my "Suzie lobbyist" outfit (suit, pantyhose, dress shoes, and briefcase). They wanted us to experience crossing the threshold of a congressional office door, to be armed with a well-thought-out list of requests, and to pitch our ideas to a young congressional aide (or, on the rare occasion, their boss) in the cramped offices of Capitol Hill. And thus, we became citizen lobbyists.

I was energized by this experience and the wider horizon offered by a national organization, so it isn't surprising that I came home longing for something beyond campus organizing for PIRGIM. After Washington Week, I made a pitch to my statewide Sierra Club chapter to be their first full-time lobbyist. There were some searching conversations about whether the chapter could afford this, but they took my proposal into consideration.

A parallel motivation was the growing excitement around land conservation in Alaska. The Alaska Native Claims Settlement Act had transferred forty-four million acres to Alaska Native ownership in 1971, but also authorized the secretary of interior to withdraw other lands for study as potential new national parks, wildlife refuges, wild and scenic rivers, and national forests. After years of planning and consultation with conservation organizations, federal agencies, Alaska's twelve newly established Native corporations, and the State of Alaska, in early 1977 Congressman Morris Udall introduced HR 39, the Alaska National Interest Lands Conservation Act. It was the largest park, wilderness, wildlife refuge, and scenic river conservation legislation in the history of the United States, and at 157 million acres, it will likely never be surpassed in scope.

Congressional field hearings were happening all over the country, and the nearest to us was in Chicago. I traveled with other Michigan Sierra Club leaders to testify at the hearing, and still have the faded paper copy of my brief handwritten testimony. I was a bit intimidated by the members of Congress behind the dais, but excited to add my voice to a rising chorus urging Congress to protect the vast federal lands of Alaska.

Composed of the nation's leading environmental organizations, the Alaska Coalition rose to become the most sophisticated and impassioned alliance in the young history of the environmental movement. I am hard-pressed to think of any other coalitional effort in the environmental movement that has equaled it since. Through its organizing moxie, it built the case for Alaska land conservation across the nation. The lands faced development from massive-scale logging, surface mining, oil extraction, and commensurate infrastructure. Many felt that Alaska was "the last, best chance" to balance conservation of globally valuable ecosystems with twentieth-century development.

I joined the coalition as a volunteer congressional district coordinator in Michigan, helping to generate citizen action to spur support by specific members of Congress. I also traveled to Washington, D.C., as the coalition began to bring waves of citizen lobbyists to Capitol Hill. Local volunteers and staff offered us places to stay—even if it was their apartment floor. It was exciting to be part of something large and historic, and it was an invaluable experience in learning the congressional process. I also met some of the brilliant strategists who understood the inner workings of Congress and its committees. They knew how to shape a legislative game plan and how to mobilize grassroots volunteers around the idea of safeguarding places most of us would never experience firsthand.

As we approached the summer of 1977, the chapter approved my proposal to be their lobbyist, and that fall, I walked into a job where the people I worked for wanted to save wild places for their own sake, and no one would give me a hard time because I hadn't skinned my own dinner.

PART TWO

To Protect the Land

What would the world be, once bereft
Of wet and of wildness? Let them be left,
O let them be left, wildness and wet;
Long live the weeds and the wilderness yet.

—From “Inversnaid,” Gerard Manley Hopkins

CHAPTER FOUR

Ah, Wilderness

In the fall of 1977, I dove into the job of chapter field representative, working on both state and federal policies related to Michigan. I continued working on the Alaska campaign while also taking on other federal land issues. Earlier in 1977, the U.S. Forest Service launched a new nationwide wilderness review called "RARE II." This stood for the Roadless Area Review and Evaluation, second time around. The idea behind RARE II was to undertake a nationwide inventory of all federally owned lands managed by the Forest Service and to assess the suitability of the roadless lands for inclusion in the federal Wilderness System.

RARE II was an attempt by the U.S. Forest Service to settle the "wilderness question" once and for all, which in retrospect seems charming and naive. To provide context for the wilderness question, we need to travel back in time.

The cultural meanings of wilderness are as complex as culture itself. Various cultures view their relationship with the natural world through different lenses, and within those lenses, viewpoints differ about places with a limited human footprint—places that we might label "natural," "wild," or "wilderness." Some embrace the human relationship with the rest of nature as kin; some view their role as caretakers and stewards. Others view the natural world with fear and suspicion, and some hardly consider nature at all. These competing cultural frameworks have led to conflicts over land and place over the centuries. Human relationships with natural and "wild" areas continue to evolve, particularly as more pluralistic cultures emerge across billions of people.

As a result, wilderness is a fraught concept. During conquest and colonization, labeling an area as "wilderness" allowed colonizers to devalue, dehumanize, or even "erase" the people already living in the landscape, relegating them to merely being part of the scenic background, and making it easier on the conscience to seize their lands. Colonists could also justify settlement as a means to conquer what they defined as "untamed wilderness" (with the implication that it needed taming) and rationalize "civilizing" these lands as an act of virtue—under their own terms, of course. Wilderness in this colonial frame carries with it heavy cultural baggage, and that baggage still affects how the continent's Indigenous peoples access their ancestral lands.

American expansion through the eighteenth and nineteenth centuries opened the door for opportunism and greed. Government policy forcefully removed Indigenous peoples from their homelands and hunting grounds as European colonists and their descendants claimed areas for their own purposes, whether that was for settlement, development, or tourism in places of natural beauty and wonder.

Historian William Cronon's critique "The Trouble with Wilderness" caused quite a stir—at least among academics, wilderness critics, and advocates—when it was first published in 1995. Wilderness opponents seized upon it to criticize the basis of the wilderness system. Proponents had mixed reactions, ranging from "this is pretty academic," to bewilderment that this made it harder to get new areas designated by Congress. Cronon argued that wilderness is peculiarly an American construct rooted in the era of western European Romanticism (1785–1832), which emphasized both "sublime" aspects of beauty and reverence for nature and a physical place to express rugged individualism as frontiers faded. Both characteristics were used to justify setting aside areas from modern development and permanent human settlement. Cronon's essay asserts: "No matter what the angle from which we regard it, wilderness offers us the illusion that we can escape the cares and troubles of the world in which our past has ensnared us."[1]

I understand these historical and cultural threads; however, concluding that wilderness is illusory escapism (beyond brief visits to refresh the spirit) may ring true for some, but not for the wilderness advocates I know, and I was uncomfortable with the broad "we" in that sentence. It wasn't what drove me, or most other wilderness proponents, to protect these areas, although I agree that many of these areas are indeed sublime. Cronon himself stresses that he wasn't arguing against wilderness protection, but rather arguing that we need to understand our human relationship with the concept and its implications.

Whether driven by romance or nostalgia, the American quest for safeguarding healthy, spectacular, or large roadless landscapes had parallel motivations to those that led to establishing national parks—protecting natural beauty and wonders from development and destruction. But wilderness protection introduced a different land management approach, more than fifty years after the first national park designation. Unlike parks, the goal of wilderness was to protect wild lands for conservation rather than the enjoyment of human tourists (although recreation was certainly among the intended uses for wilderness). Biology and the emerging science of ecology helped drive the wilderness movement.

Wilderness with a capital "W" emerged in the twentieth century as a formal designation restricting development and other specific activities on lands already within federal public ownership. The much-celebrated conservationist and writer Aldo Leopold was a pivotal early figure in the U.S. wilderness story. As a young Forest

Service employee, he witnessed the impacts of mining, logging, and ranching on remote and rugged backcountry areas. He understood these threats before most and realized the need to protect these areas to support the health of the land, water, and wildlife. He became an early advocate within the Forest Service for setting aside large areas where natural systems could function free from resource extraction and still be available for backcountry recreation. In response, the Forest Service designated the country's first formal wilderness—more than five hundred thousand acres of mountains, rivers, and desert surrounding the Gila River—as the Gila Wilderness in 1924.

As the young supervisor of New Mexico's Carson National Forest in 1912, Leopold's duties included killing wolves, on the premise that fewer predators meant more game for human hunters.[2] In his seminal work *A Sand County Almanac*, he writes of one particular moment after taking down a wolf.

> I realized then, and have known ever since, that there was something new to me in those eyes—something known only to her and to the mountain. I was young then, and full of trigger-itch; I thought that because fewer wolves meant more deer, that no wolves would mean hunters' paradise. But after seeing the green fire die, I sensed that neither the wolf nor the mountain agreed with such a view.[3]

His "Thinking Like a Mountain" essay emerged from this experience and reflects a turning point in his views on conservation. He began to think about land and wildlife management from the perspective of whole ecosystems rather than an effort to promote human-favored species. As the "mountain" knows, a one-size-fits-all policy based on human assumptions about how nature functions doesn't work for complex ecosystems. A healthy ecosystem needs its predators. This was radical thinking for a young graduate of the Yale School of Forestry in the early twentieth century.

Leopold soon connected with a handful of other like-minded leaders, and they founded the Wilderness Society in 1935. As noted on their website, "At the time, forests and other public lands were mainly seen as resources for logging, mining, and other development. Setting out to conserve the wildest places was a revolutionary idea."[4] Cofounder Bob Marshall, also a forester by training with a PhD in plant physiology, represents, like Leopold, an important thread in the wilderness movement, in which biologists, foresters, ecologists, and other scientists played key roles as advocates for protecting natural systems. While serving as director of the Indian Forest Service, Marshall also championed the rights of Native Americans within this federal agency and proposed subsidized travel to enable families with limited means to experience the beauty of the wilderness—an idea that has resurfaced in the 2020s. He died young at thirty-nine, but his passion for wilderness protection left a legacy carried forward by his friends and colleagues.

After World War II, industrial growth and development accelerated in the United States, bringing a surge of new highways, cars, and houses. In the same era, logging rates increased in the national forests, which provided substantial lumber for new houses and pulp for the paper industries. Public lands also provided grazing lands for cattle, as well as places to hunt, camp, and play. More timber extraction required more roads. In 1944, Smokey Bear became a Forest Service symbol to promote forest fire prevention, because we needed these forests. The agency embedded "Only YOU Can Prevent Forest Fires!" in the memories of the early postwar generations.

Botanist and conservation biologist Don Waller pointed out the double irony of this slogan: "Fires were a natural and integral part of many of our western federal forestlands where fire suppression (and often logging) have greatly increased risks of more catastrophic fires."[5] The agencies making the case to protect forests were at the same time purposefully carving up the vast forested landscapes of the American West, with more logging, including ancient forests and on steep slopes.

In the 1950s, wilderness champions were shaping strategies to create a federal framework for protecting wild and natural places. The Bureau of Reclamation planned to build a hydroelectric dam at Echo Park within Dinosaur National Monument, sparking nationwide opposition. The Sierra Club leaned in to oppose the dam, joined by other park and wilderness advocates in the Council of Conservationists—a group of nine organizations that included the Wilderness Society and Audubon Society. The coalitional effort mobilized the membership of these groups and served as a model for future campaign strategies in the coming decades. Essays, news features, films, and many letters to Congress from constituents all over the United States built the case for protecting the monument and its irreplaceable fossils and artifacts. In 1956, Congress removed Echo Park Dam from the legislation to authorize the Colorado Basin Storage Project, and conservationists had their victory.

That same year, Howard Zahniser, executive secretary of the Wilderness Society, wrote the first draft of something he called the "Wilderness Act." A journalist and former staffer for the U.S. Biological Survey, he brought both writing skills and literal field experience to his task. The Wilderness Society notes that "the bill would be rewritten or resubmitted 66 times, subject to 18 public hearings and 16,000 pages of testimony."[6] It took tenacity; it also took eight years to enact. All things considered, that's a short timeline for what it takes to get an idea for something as monumental as a new federal land management system through the legislative process—and that was with a functioning bipartisan Congress.

After all his efforts, Zahniser died just a few months before President Johnson signed the 1964 Wilderness Act into law. It codified the Gila Wilderness, and added fifty-three other new areas, including the Bob Marshall Wilderness in Montana and the Boundary Waters Canoe Area Wilderness in Minnesota. It also opened the door for future designations by Congress. As of 2022, the system Zahniser envisioned includes eight hundred designated Wilderness areas managed by four federal agencies that protect over 111 million acres of land and water in the United States.[7]

The Wilderness Act spanned all the federal land management agencies. While most Wilderness designations have been in national forests, there are also Wilderness areas in national wildlife refuges, national parks, and Bureau of Land Management lands. These distinct agencies manage lands under different mandates and priorities, but regardless of the agency, they manage Wilderness areas to protect their intrinsic and ecological value.

In all those drafts, Zahniser had the unenviable job of coming up with a definition for what was and wasn't "wilderness" for the United States of America. His final description was as follows: "A wilderness, in contrast with those areas where man and his own works dominate the landscape, is hereby recognized as an area where the earth and its community of life are untrammeled by man, where man himself is a visitor who does not remain."[8] This eloquent passage was both strategically brilliant within its political moment and the source of much argument over interpretations in subsequent years. "Untrammeled" is a word that most of us don't use frequently, and some have incorrectly substituted the word "untrampled" in their interpretation.

Doug Scott, a wilderness champion in his own right (and one of my former department heads at the Sierra Club), wrote about that highly charged word "untrammeled" in a 2001 article in *Wild Earth*. "Current dictionary descriptions of the word . . . include 'unrestrained,' 'unrestricted,' 'unimpeded,' 'unencumbered,' 'unconfined,' 'unlimited.'" Scott points out, "the [Wilderness] act applies the word untrammeled not to an 'area' or its present condition, but to 'the earth and its community of life,' that is, to the forces of Nature." So we can think of wilderness as where the forces of Nature are not impeded. Congress added a second sentence in their definition, spelling out what Zahniser described as the practical applications within wilderness, He maintained that the first sentence characterizes the ideal of wilderness.[9] Today, distant from the politics of that moment, "untrammeled" (like "wilderness") is defined by some in ways that erase or deny Native people's presence and continuing traditional uses of their lands, but that wasn't the original aim, nor is it practice in Wilderness management.

Within the larger wilderness debate, there has been considerable argument over terms such as "pristine" wilderness, and the degree to which a given area had been previously disturbed or modified by humans, although the concept no longer has much value or traction. In the early years of the wilderness movement, reluctant federal agency staff and those simply opposed to protecting wild areas used the "pristine" argument to keep areas out of the federal Wilderness system.

Some argued that historic logging or evidence of an old road track should disqualify an area—often the case for those hung up on the notion of "virgin" wilderness. Arguing about what is pristine enough to qualify as wilderness seems a moot point in our current era when various human-made contaminants and climate change extend to every corner of the globe from Antarctic penguins to the albatrosses nesting on the remote islands of the Midway Atoll. We now live in an era that geologists term the Anthropocene in recognition of how dominant human activities have become. What corner of the biosphere lacks human influence?

A different argument is that Indigenous peoples have manipulated North American landscapes for thousands of years—via practices like prairie burning, raising food crops, and selective logging—making a hands-off approach to managing wilderness artificial. Most Wilderness areas allow some hunting, fishing, foraging, and prescribed burns, but one can argue that the very light touch of human management in Wilderness does not mirror historical practice. Then again, today's population density of 1.3 billion people in the hemisphere hardly compares with the perhaps 60 million people estimated to have occupied the Americas before 1492. For those Indigenous peoples, land management involved careful balance and relationships to yield sustainable supplies of food, shelter, and goods from healthy ecosystems. It is safe to say that current practices fail to emulate those sustainable practices on public and private lands across many American landscapes. But that doesn't mean it has to stay this way.

The fact that nothing on the surface of Earth and its atmosphere is unaltered by the hand of humankind doesn't justify manipulating and altering everything else. With so much human demand on natural systems today, doesn't it make sense to provide some breathing room on the landscape? That was part of the original impetus beyond the romance of wilderness. Human survival depends on healthy, functioning, natural systems, as does all life on earth, and letting nature do her work for the living biosphere where she can do it without a lot of disruptive human interference is an act of mutual respect and survival.

The scientific community is arguing for a dramatic increase in protected conservation lands—a concept that embraces a wide definition of "protected" and "conservation"—as an important survival strategy in the face of climate change. In 2023, the International Union for the Conservation of Nature (IUCN) estimated that about 15 percent of terrestrial earth and only about 7.5 percent of the oceans have some protected status.[10] That's not enough.

IUCN is leading a global effort to increase conservation land, asserting: "We need these conserved areas to be vital signs of life . . . to offer hope, regeneration, and good health. We depend on them for clean air, fresh water and rainfall, pollination, and spiritual sanctuary to people all over the world, in urban and in rural communities."[11] Complementary efforts include "Nature Needs Half," launched by a coalition of scientists, conservation groups, and Indigenous leaders with the bold goal of protecting biodiversity and buffering climate change through conserving half of all the earth's 846 ecoregions by 2030. A parallel effort, the Global Deal for Nature calls for 30 percent of terrestrial habitats to be formally protected and an additional 20 percent designated as climate stabilization areas by 2030.[12]

These new initiatives stress working with and for resident Indigenous and other land-dependent people—the goal is ecologically healthy lands that support life on Earth, including our lives. A planetary-scale ecological safety net might not have been the primary impetus for protecting undeveloped land in the early years of the wilderness movement, but it seems the most compelling argument in today's world.

So, we have multiple cultural constructs about what we think wilderness *is*, with or without that capital "W," and why it does or does not matter in our pluralistic society. We also have a federal legal framework born out of some of those constructs with artfully chosen words that defines what the law thinks Wilderness is. These shape, but don't necessarily determine, the outcome of a complicated democratic process that further influences what will happen on the land. Somehow, out of all of this, we humans need to figure out our relationship with and within the natural world, in ways that are mutually beneficial. In the age of the Anthropocene, will we strive to work this out? The IUCN strategy is promising. Here in the United States, another hopeful approach is comanagement of public lands—formal management partnerships between tribal governments and the federal government for public lands. The National Park Service already has several agreements in place, and many more are anticipated.

The evolution of public lands and their management in the United States, including wilderness and parks, has been at times clumsy, grossly insensitive to the Indigenous peoples on this continent, and shaped by the dominant cultures and their values within various eras. It can inform, but doesn't necessarily constrain, future strategies. At the same time, this imperfect system has created a collection of remnant spaces on a human-dominated planet in which nature in her damaged but tenacious elegance can survive, if not thrive, for now. Surely, we owe our mother planet at least that much.

CHAPTER FIVE

Welcome to the Big Leagues

Prompted by the passage of the Wilderness Act of 1964, the Forest Service embarked on an initial inventory to assess potential Wilderness areas on national forest lands. This first Roadless Area Review and Evaluation was known as RARE I.

Across most of the forty-eight contiguous U.S. states, you are never very far from a road. The presence or absence of roads served as the primary indicator of "development" within these lands for this, and the next evaluation. In the contiguous states, the place furthest from a road—and therefore the remotest spot—is the Thorofare Wilderness in the heart of Yellowstone National Park in Wyoming, which is 18.7 miles from the nearest road as the crow flies and a seventy-five-mile walk via trail.[1]

The RARE I study primarily focused on large tracts in the American West. Not surprisingly, when completed in 1972, there were few recommendations for Wilderness designations in the Midwest or Eastern states. This led to public clamor for a more thorough inventory that included the cut-over and regrown forests of the

East. Ernie Dickerman, a lifelong wilderness champion, helped lead the first big effort that culminated in the Eastern Wilderness Act in 1975. It was a start. It added sixteen areas to the Wilderness System, only one of which was in the Midwest: the Rainbow Lakes area in Wisconsin. While a remarkable achievement, many felt it was a drop in the bucket and continued to push for more opportunities in the East. In 1977, the Forest Service initiated another evaluation: RARE II.

The timing coincided with my new job with the Mackinac (Michigan) Chapter, and RARE II became a primary focus. I juggled this with assignments in the state legislature, the most significant of which was legislation to protect Michigan wetlands. I also worked on air-quality standards, sand dune protection, watershed planning, revoking the coyote bounty, and many other issues that fit under the broad environmental umbrella. The movement wasn't deeply specialized back then, and it was common to move from land and water issues to health and energy in any given week—all part of a larger fabric of a healthy environment.

Early in the RARE II process, the Forest Service provided an opportunity for citizens to recommend areas for evaluation. One of Michigan's most seasoned advocates for wild places, Virginia Prentice, stepped up to help identify candidate areas in Michigan.

Virginia was the classic little old lady in tennis shoes, but what a powerhouse. She had already paid her dues in the long, heated battles to establish Pictured Rocks (in 1966) and Sleeping Bear Dunes (in 1970) National Lakeshores and had championed wilderness protection for the interior of Isle Royale National Park. She knew the rules of the land management game and the federal policy process, and she knew Michigan's lands in a way few others did. She pored over Forest Service maps with Steve Payne, the Midwestern staffer from the Wilderness Society, based in Duluth. They had additional input from U.P. volunteers, including Chuck Meyer and Roger Harbin, who knew the forests well.

Together, this informal team circled a dozen or so promising areas in the vast acreage of Michigan's national forests, ranging from a few tiny Great Lakes islands to tracts of nearly twenty thousand acres in the western U.P. This became the initial template from which the

Sierra Club's proposal for Michigan Wilderness designations would evolve. Later, opposing groups would disparage this effort as just a bunch of amateurs drawing circles on a map. But "amateur" is a word that describes one who does something for love, and Virginia, as well as her friend Cindy Thomson, and Steve, Chuck, and Roger, in addition to so many others who joined in this effort, loved the wild forests and shores of Michigan. I find this qualification enough to begin a public discussion about public land.

As noted, roads crisscross the U.S. landscape at such a density that it isn't surprising that the Forest Service's own inventory identified many of the same areas, distinguished primarily by the absence of roads within them.[2] Many of these areas had been spared roads because of their steep topography, including the Sturgeon River Gorge and Rock River Canyon. Some were undeveloped because they were huge swamps, like the place named "Delirium" for the effect of its black fly and mosquito population on trappers. Others were lucky flukes.

The largest two areas were wild and largely intact because they had been in private holdings for the better part of a century. One was Cyrus McCormick's seventeen-thousand-acre summer retreat south of the Huron Mountains, and the other the twenty-thousand-acre Sylvania Tract, a private hunting and fishing club on the Michigan-Wisconsin border. Both these tracts had escaped the logging boom, and descendants of the original owners had recently sold or deeded them to the Forest Service. The Forest Service excluded the McCormick Tract from their recommendations because they were managing it for research on primitive recreation and the wilderness experience. We thought this was perfectly compatible with a Wilderness designation; their science team did not. Undeterred, we recommended McCormick in our legislative proposal anyway, appealing to Rupert Cutler, a Michigander serving as assistant secretary of agriculture for conservation, research, and education, who was leading the agency's RARE II effort.

Seasoned colleagues who had succeeded in persuading Congress to enact other Wilderness bills told us we needed a big coalition of diverse state and local organizations to win. They also told us we needed at least some vocal support in the congressional districts where the proposed Wilderness areas were located. Except for Nordhouse Dunes along Lake Michigan's eastern shore, and Bear Swamp in the Huron-Manistee National Forest, all the prospective areas were in the U.P. and in one congressional district. The Sierra Club was not widely loved in the U.P., and we knew we needed allies in the north country.

Like many other states that had battled over federally designated Wilderness, in Michigan the most enthusiastic support came from urban areas and college towns, and the greatest resistance came from the communities close to and sometimes surrounded by the national forests. Many factors led to this resistance, including resentment of federal ownership and control of local lands, in addition to the lack of taxable assets on those lands. The latter poses a particular challenge for funding schools in rural forest communities in states that depend on property taxes for school support, which includes Michigan. There is also a sense from local residents that "these are our forests, and we don't need outsiders telling us what we can or cannot do on them." Add to this the attitude "Lansing and Washington don't care about us" that many in the U.P. still feel, and you can understand why the Wilderness proposal wasn't met with marching bands and bouquets in the U.P.

We hosted a Great Lakes wilderness conference at MSU in the spring of 1978 to raise awareness and build support. The attendees were, by and large, enthusiastic, but among them were two staunch opponents from the U.P., one of whom threatened to meet us with a shotgun if we ever set foot near his property. It leaves an impression.

Given their clout in the state, we sought the support of MUCC, which had affiliated local clubs and members throughout Michigan. We met with Tom Washington, and his initial objection had been the inclusion of Bear Swamp, which he felt we should drop from the package. Among other things, he wanted bear hunters to be able to haul their kills out by vehicles, and designation would prohibit motorized vehicles in the area. Believing at the time that MUCC's support was essential, we agreed to drop Bear Swamp from the proposal. It was a small area and less likely to be developed. It seemed like the best strategic choice at the time. We should have known the deal wasn't going to be this easy.

As an affiliate member, Sierra Club had a seat as a delegate and a vote at MUCC's annual meetings. They ran these big gatherings something like a political convention combined with a hunting lodge reunion. It was, in those days, predominantly male, although there were "ladies' activities" organized for the spouses of the delegates. The active presumption was that none of the delegates were women and none of the spouses were male. We decided to take our case to the upcoming annual convention in June 1978.

The Sierra Club and allies floated a resolution of support for federal Wilderness designation, and it was up for a vote by the full body, which included perhaps a hundred or more delegates from around the state. A handful of us—from Sierra Club, the regional Environmental Action Councils, Audubon, and the state's environmental education association—formed the core block of wilderness proponents at the convention. Our outreach to the other delegates and an informal head count indicated that we had a good chance of securing a Wilderness endorsement.

When our resolution came up, the debate was going fairly well. Then a senior delegate suddenly called for a recess, which the presiding officer agreed to allow. A large block of delegates got up and followed the man who had called for a break in the debate. They marched directly into the men's bathroom and closed the door, and I knew something was up.

When the assembly reconvened, the debate shifted against us. Someone from the bathroom caucus called the question and the assembly defeated our resolution. It gave new meaning to "closed-door politics" and left Tom Washington without a formal position on wilderness—something he wanted, as the congressional delegation and other leaders would seek MUCC's input. He worked with the board president to set up a committee (which included me) to work out a position statement that the board would consider a few months later. The result was a brief and neutral statement along the lines of "wilderness is a legitimate use within the multiple uses for national forests," but it fell short of engaging directly with our proposal for congressional action. While MUCC didn't support us, they didn't put much energy into opposing us either, which I considered something of a consolation prize. With or without their support, we forged ahead.

Most of the proposed Wilderness areas were remote and not well known. Following our spring conference, we organized a crew of volunteers with plans to walk through each of the prospective sites, take notes and photographs, identify any potential concerns, and develop further recommendations based on the site visit.

Two volunteers, Ken Smith and Jeff Brenner, signed up for the two-week blitz. Borrowing a tactical page from those who had used Ansel Adams photographs to showcase the need to protect spectacular areas in the West, we also brought along a team of skilled, large-format photographers who volunteered their time to document each area. In late June, our little expedition headed to the U.P., which we traversed from east to west, noting highlights, checking boundaries, being swarmed by mosquitoes, getting covered with ticks, occasionally getting lost, and helping photographers Chuck and Edna Elfont haul their gear up ridges, down ravines, and into swamps to capture the beauty and intrigue of these areas.

Taking expensive equipment into the wilds has its hazards, as any nature photographer will affirm. One afternoon, Chuck was trying to get a good shot in the Big Island Lakes area in the central U.P. He didn't realize that he had set up his tripod and camera on the edge of a floating bog mat. Shoreline vegetation around bogs can extend over the water on dense mats of sphagnum moss, supporting, at times, the considerable weight of roots and other vegetation—even trees—on the water. Chuck was concentrating on getting the perfect light and framing for his photo, as he and his tripod slowly began to sink into the bog mat. As soon as he realized what was happening, he captured the shot and quickly made his way back to drier land, with wet boots, but dry equipment in tow.

Sylvania was among our last stops on this tour of ten sites, and by the time we were at area number nine, the group dynamic was fraying a bit at the edges. Ken Smith was seeking space and time away from the group, so I agreed to hike into a backcountry campsite with him at Sylvania and spend the night in a tent rather than hang out with the rest of the crew in a motel. It poured cats and dogs that night and the mosquitoes were merciless, but there we were. The wilderness experience can lose a bit of its mystique when your focus shifts to the unpleasantness of damp socks, but symbolically we felt good about camping in a prospective Wilderness area.

The next day, the skies had cleared and the rest of the team—clean and dry—arrived refreshed for a day of scouting and photography. Sylvania gave us her best side that day, and I can still picture Chuck's gorgeous black-and-white shots of the tall trees that rim Golden Silence Lake and the craggy little islands and ancient white pines of Clark Lake. We used these images in newsletters and press packets and every other way we could think of to draw attention to Sylvania and the other candidate areas.

We headed home with high hopes for Michigan wilderness, not knowing it would take countless meetings with potential allies, members of Congress, committee staff, editorial boards, property owners, Forest Service staff, and others to make it down the long and winding road through two houses of Congress and the final step of securing the president's signature. The MUCC disappointment was only one of the many hurdles along the way. I don't think we imagined in the early years of this struggle that we would be in for a full decade of uphill climbs. We focused on immediate challenges, and one of those was finding enough money to keep a staffed office going.

In 1979, our two-year experiment with funding a full-time chapter lobbyist was hitting the tough realities of a gap between income and expenses. I volunteered to go off payroll for part of the summer to work on a proposed book project on the Michigan backcountry trails. I covered a lot of territory in those weeks, often solo, sometimes with friends. My boots and backpack took me to a private moonlit perch on top of Nordhouse Dunes, along trails with amiable hiking companion Cheryl Bartz in the verdant Jordan River valley, and across stretches of the early stages of the North Country Trail with Sierra Club pal Ken Magnell.

My treks included a multiday solo trip in Pictured Rocks, which coincided with the July presence of stable flies. These little beasties look like innocent houseflies, but they want your flesh, and they will help themselves to it without hesitation. Hot, humid days attract them to Lake Superior shores, which is exactly when an over-heated, tired human wants to slip into the water and enjoy the sandy beach. Wet human skin attracts them like a magnet. On my second day, I made camp at the spectacular Chapel Beach campsite and had the entire place to myself, except for the squadrons of stable flies. I set aside any notions about wilderness enlightenment in this gorgeous setting and focused on the urgent need to cope with the heat, humidity, and flies.

I knew that only a fool would swim alone in Lake Superior when she was tired, hot, and far from any assistance. The tent offered relief from the flies but was a nylon sauna. I finally resorted to pulling on my mosquito head net and my swimsuit and found a shallow place in the Chapel River where I could lie down covered by the cool rushing water with my head propped up on a rock. It offered temporary but much-needed relief. When I told friends and family stories about experiences like this, a common response was: "And you do this for fun?"

Sleeping Bear Dunes was a gentler, if more populated, experience. I set up housekeeping in the Platte River campground with a full neighborhood of car campers. It was cherry season, and on a side trip I picked up a quart at a highway stand. Toward the end of the day, I headed down to the mouth of the Platte River to watch the sunset. Others had the same idea, and families were lingering on the beach after an afternoon in the sun.

The westward-flowing Platte takes a right-angle turn just before it meets Lake Michigan and runs for a few hundred yards between a sandy beach and offshore sandbar before it merges with the lake. In the right conditions, it's an easy wade out to the sandbar, and I headed out there with my camera, notebook, and quart of cherries. The air was warm and moist in the height of summer, and the lake was rolling in easy while children laughed and played on the shoreline. A handful of other campers forded the Platte and joined me on the sandbar, coming out with wet shorts and laughter. We struck up an easy conversation as we anticipated the light show to come.

I passed the cherries around, which were sun-warmed, and exquisitely fresh. My new companions and I talked about how lovely this place was, how great summer was on the shores of Lake Michigan, where people were from, and how we'd found this river.

The sun began its descent into the lake and became that bright-orange ball that you always expect to sizzle when it hits the water. The ball distorts at a point where it is about halfway beyond the horizon, and then its crescent slips silently below the waterline. The afterglow in a cloudless sky progresses into an apricot blush, then rose-pink, followed by the gray tones in the pink that fade into the deep blues and purples of twilight, and then night embraces the lake. Sometimes, life is absolutely perfect. That sunset shared with gentle strangers on the shores of Platte Bay was one of those moments.

By fall, it was clear that the chapter didn't have the resources to keep me on for more than eight weeks. While the chapter executive committee met to work out the details in the main office space (under leaky skylights, where pigeon feathers would float down to our desks in our low-rent "attic" office), I retreated to a private office and indulged in a good gut-wrenching sob. Sometimes there's absolutely no value in sucking it up and putting on a good face, and an honest heartfelt weep is the easiest way to acknowledge grief and disappointment so you can get it out and move forward. After I mopped up my tears, we made plans for what the chapter would do next as I tried to figure out the same for myself.

Sometimes, too, another door opens when you need it the most and expect it the least. One of the field representatives at the Sierra Club's Midwest office in Madison, Wisconsin, had decided to return to her former home in Seattle, and the position was open. I'd been a finalist in their previous search, so I was optimistic about a second chance. While I was on my flight to Madison for the interview, the woman sitting next to me extolled the virtues of Madison. She said, "If you get this job, you'll never want to leave." The senior field representative offered me the position, and I quickly accepted. A few weeks later, I said farewell to Michigan, packed up a little U-Haul, and headed off to a new place to call home. In early December 1979, I walked into the regional office in Madison. I carried my wilderness agenda with me to a larger base of operations and joined a team of some of the best and brightest in the nation's environmental movement.

When I joined the national staff of the Sierra Club, the field staff were dispersed around the country in ten or so regional offices. I was thrilled to have the opportunity to be part of this smart and intensely committed team. We were lobbyists, organizers, issue experts, and media spokespeople, shifting from one role to another at the drop of a hat. We also served as political strategists for legislative and regulatory policy. This demanded skill, instinct, and, for survival, a robust sense of humor.

The Midwest office had nine states in its territory, and we covered the workload with two field representatives and an administrative assistant. We were responsible for the largest block of Congress in the entire field system—roughly one hundred members of the House and Senate. Our assignments focused primarily on federal issues, although we also worked on major regional issues that cut across state lines. (State issues were typically the territory of the statewide chapters.)

My early portfolio included a final push to secure the Alaska lands bill, the Coastal Zone Management Act, the Michigan, Indiana, and Wisconsin Wilderness bills, and whatever my boss, or the leadership teams in Washington, D.C., and San Francisco, sent my way. While the Michigan Wilderness bill was a priority, it had to share space in my workload with everything else. Over my nearly twelve years on the national field staff, I worked on other Wilderness bills across the country and national park and national lakeshore legislation, including funding for operations and land acquisition. We carried significant lobbying assignments on environmental quality bills including the 1986 reauthorization of Superfund (the legislation to fund cleanup of toxic hot spots) and the Midwest Compact (for low-level radioactive waste management). We played a major substantive role in the 1987 Clean Water Act Amendments, the 1990 Clean Air Act Amendments, and the Great Lakes Critical Programs Act of 1990, as well as assignments on countless other bills. While my major efforts were on public lands, and work related to the Great Lakes, we had to be up to speed and literate on all current priorities whenever San Francisco or Washington called us to action.

We slogged to get the Superfund amendments passed in the late 1980s. The nation was reckoning with the toxic waste at places like Love Canal in New York, and Times Beach in Missouri, and how to find the funds to clean up these sites. I knew my stuff on wilderness and water, but making the case for something like joint and several liability on abandoned toxic waste sites was not my forte, and the corporate lawyers dominated the playing field. We would get urgent emails calling for "all hands on deck" and would put other priorities on hold or scrap vacation plans to rise to the call.

Sometimes our survival humor became a bit cynical, but it helped keep the team going on the long-haul issues. This was expressed through parodied versions of newsletters, magazine features, and management directives, as well as occasional evening skits during multiday meetings. The latter grew into staged parodies of Broadway musicals that we produced at annual meetings in San Francisco for many years, from which "Greener than Greenpeace" to the tune of Richard Rogers's "Younger Than Springtime" was one of my brief hits. During lobby weeks, there were often late-night sessions to blow off steam at Washington's Tabard Inn, and an annual covert ceremony of alternative staff awards (one of which was sufficiently boisterous to compel the Tabard to request that we never again hold this event on their premises). When we finally passed the 1986 amendments, I was in a staff skit where my line was, "As God as my witness, I will never work Superfund again."

I was keenly aware of my privilege to work with brilliant environmental strategists in the policymaking arena. They understood process, power, and opportunity, including when to apply pressure and when to back off. They understood the value of trusted relationships with allies, and, before "message" was a verb, how to effectively communicate to rally the grassroots and the media. We among the younger staff were always absorbing and learning. All of us worked at an extraordinary pace and intensity, moving from one urgent matter to the next. One of my quotes from the Sierra Club years is "There is always a crisis waiting to fill any vacuum."

So, in my new hometown (the woman on the plane was right), I was working with this winning team, earning a modest living, and life was good. It was exciting and rewarding but often exhausting and sometimes crushing. I took our work very seriously because the outcomes affected so many people and places and our very quality of life. I'm still not over some of the losses.

One evening, in my early years on the national staff, I paused on the Pennsylvania Avenue sidewalk, looking at the Capitol building illuminated in the dark night. I turned to my companion, John McComb, who would soon become the head of our Washington, D.C., office, and asked: "Do you ever think about it—do you ever look at all this and say, 'I'm part of this; I'm part of making policy that could affect millions of lives and the future?'" He smiled at me with that "You're new at this, aren't you?" look on his face, and the experience of someone who had seen far more of the gritty deal-making of policy than I. He replied, "I don't think about it that way very often, but it's good to be part of making things happen." And it *was* good, but sometimes so hard and discouraging. Tenacity and perseverance in this work are essential.

Ultimately, for Michigan Wilderness, it would take another eight years through five sequential congressional sessions to achieve victory from the time I joined the Midwest office. It would also take what Brock Evans always described as an essential strategy to move policy forward: "endless pressure, endlessly applied."

We first got traction when Representative Bob Carr from Lansing introduced a bill in the House of Representatives, and Senator Donald Riegle introduced a Senate version in early 1980. The legislation called for protecting eleven areas, comprising slightly more than ninety thousand acres. The largest areas were just over eighteen thousand acres of Sylvania, and the seventeen-thousand-acre McCormick Tract, both in the western U.P.'s Ottawa National Forest. The smallest were two Great Lakes islands managed by the Hiawatha National Forest—Government Island in the Les Cheneaux chain in northeastern Lake Huron and Round Island, in the Straits of Mackinac—each only a few hundred acres.

The Mackinac Chapter had been rebuilding its resources and recruited new staff, and in the early 1980s, Sue Pemberton became my Michigan partner in the legislative strategy. Our legislative momentum was rapidly quelled when Senator Riegle got very cold feet after wilderness opponents confronted him at a late October meeting in Escanaba that same year.

Their arguments weren't new. What about valuable timber? What about managing for game species? I won't be able to drive my vehicle in these areas. The senator wasn't prepared to offer the counterarguments about the multiple uses of national forest lands (which include wilderness), or that people could still hunt and fish, or that some people really enjoy primitive recreation away from cars and clear-cuts, and that a few tiny areas in the state that allowed this shouldn't be such a hardship. He rapidly backed away from the legislation. A few days later, Bob Carr lost his seat in the 1980 election. Our fortunes had changed.

In the summer of 1981, we hosted another Wilderness tour, joined by one of Bob Carr's former staffers, Dave Dempsey. That same year, Congressman Dale Kildee from Flint secured a seat on the House Committee on Interior and Insular Affairs, a critical committee for wilderness and park deliberations. Kildee stepped into the role of champion, and introduced a new Michigan Wilderness bill, H.R. 148, in the 97th Congress, and each succeeding Congress through the final passage in the 100th Congress, but the road was uphill without a companion Senate bill.

As we moved into 1984, we were watching Congress pass bills in other states—twenty-one, in fact—and there was a sense that we were falling behind. The staff and volunteer team rallied to continue to press Congress. That summer, we organized another congressional field trip, working primarily with Senate staff and Forest Service personnel. Over three days in August, we took them to Sylvania, McCormick, Sturgeon River Gorge, Rock River Canyon, and Big Island Lakes, and included an extended loop over to Pictured Rocks on the final day to squeeze in as much federal lands background in one trip as we could. It is so much easier to advocate for the value of these areas when people charged with the tough decisions to protect them can see them firsthand and experience their wonder. We wrapped up with an 11 p.m. meeting with U.P. wilderness volunteers at the Marquette Holiday Inn on the last night, weary but hopeful.

We entered the 99th Congress in 1985 with renewed hope, buoyed by another big wilderness conference in Ann Arbor at the end of January. Anne Woiwode succeeded Sue Pemberton as the Mackinac Chapter director, and jumped right in. Over the coming months, we secured ten House cosponsors—two Republicans and eight Democrats (including Bob Carr, who was back in Congress). In April, the Michigan Forest Products Industry Development Council endorsed H.R. 148. Then, Governor Blanchard, encouraged by his environmental advisor Dave Dempsey, endorsed the legislation at the Michigan Chapter's annual banquet in Ann Arbor.[3]

We had momentum, but it was fragile. Our champions had scheduled H.R. 148 for hearings in the House, and Senator Levin's staff was warming up to the idea of a Senate bill. We continued to keep the lines of communication open and active with Riegle's office, as we needed at least his tacit support on not only the Wilderness Bill but also commitments on other environmental bills moving through Congress.

On June 25, 1985, the House Committee on Interior and Insular Affairs Subcommittee on Public Lands held hearings on Kildee's H.R. 148. The Michigan Chapter sent more than a dozen volunteers to testify and lobby their members of Congress, and Anne and I joined them making the rounds on Capitol Hill and at the hearing table. Kildee felt it was important for him to listen to people in the U.P. and traveled with committee staff for an informal meeting a few weeks later. Once again, the Marquette Holiday Inn played a role in advancing the dialogue. (I propose a plaque in the lobby to commemorate its pivotal role as a neutral convening site.) It was tense, but Kildee patiently listened to testimony for several hours, and we were both surprised and pleased by the number of supporters who spoke up on behalf of the bill. We knew they were out there, and we were heartened to see fresh faces at the table.

The bill passed in the House of Representatives for the first time on September 17, 1985—an extraordinary step forward. The Senate was still a dilemma, as Anne recalled:

> Under Congressional rules, a bill needed someone to introduce a Senate bill for the superior house to consider the matter, and tradition held that the Wilderness bill sponsor needed to be a Michigander. Senator Riegle's very public about-face meant his colleague Senator Carl Levin was on the spot.[4]

Senator Levin introduced his bill shortly after the House Bill passed, but it excluded Delirium and Government Island. We all knew that we needed Riegle to refrain from actively opposing the bill to avoid splitting Democratic votes in the Senate. We continued to meet with his staff and kept the grassroots letters coming. We also had the impression that before bringing a bill forward, Senator Levin was working behind the scenes with Senator Riegle to gain his colleague's quiet acquiescence.

Levin's bill had little attention until the fall of 1986. Just when it looked like the stars were aligning to succeed in the 99th Congress, enter the oil and gas industry, stage right. Many people aren't aware that Michigan has oil and gas wells, but the state's underlying geological formations have oil and gas deposits, especially within the Niagaran reef underneath the arc of the Lower Peninsula.

Many of the mineral rights under the National Forests in the eastern United States are privately owned, as was the case under the sands and forests of Nordhouse Dunes. The rightsholders didn't want the complicating factor of a Wilderness designation. While the oil and gas companies could have accessed these potential sources by a process called "slant drilling" that would have kept rigs out of the proposed Wilderness area, they sought a simpler solution. Enter, from stage farther right, the Reagan administration's Washington, D.C., office of the U.S. Forest Service. In the late summer of 1986, just when the Senate was getting close to moving on Senator Levin's bill, the Forest Service stated (with almost no warning to their people in the field, the wilderness advocates, or Senator Levin) that they would oppose the bill if it included Nordhouse Dunes. We'd been thrown under the bus.

The Senate Committee on Agriculture (the Forest Service is part of the USDA) was holding the hearing on Levin's Bill. Committee Chair Jesse Helms was absent. The next-ranking member had to leave early, and asked Riegle (not even a member of the committee) to chair the rest of the hearing, as he was the highest-ranking member of the Senate in the room. We weren't the only ones in shock. This bizarre parliamentary procedure placed Riegle (who really didn't want any of the spotlight on this highly charged issue) smack in the middle of a most awkward decision. The Forest Service delivered their testimony, stating their new opposition to Nordhouse, and as the hearing closed, we knew the bill was dead in the water for the 99th Congress. Following the hearing, I headed back to the club's D.C. office and called the head of the Forest Service in Region 9 back in Milwaukee and let it rip. I do not yell at people over the phone—ever—but I did that day. Anne was a witness, and I trust her recollection. She sent me out for a walk to cool off. It worked in the moment, but I was devastated, worried that we might have lost our shot.

In the aftermath of the bait-and-switch hearing from hell, Senator Levin was nearly as outraged as we were that the Reagan administration had pulled this last-minute maneuver. It galvanized his commitment, and he was emphatic about not sacrificing Nordhouse Dunes and not giving up. Still, we had to wait for the 100th Congress to start all over again in 1987. As Anne noted:

> Congressman Kildee and Senator Levin both moved quickly to introduce their respective bills in the first months of 1987, not willing to face the kind of last-minute tricks played by the Forest Service the year before. Both bills worked their way through hearings and the Forest Service continued to oppose designation of Nordhouse Dunes despite reassurances on mineral rights issues.[5]

Kildee and Levin (and probably Riegle as well) were as eager as we to wrap this up and move on. Strong support from the majority of the Michigan delegation helped the House bill pass early in the year. Later in the fall, Senator Levin's bill moved easily through the committee with no drama this time. Levin and Kildee worked through their minor differences, and agreed to substitute the House version on the Senate floor, sans Government Island (thus avoiding any possible disputes in a conference committee). The Senate acted quickly thereafter, and on December 8, 1987, President Reagan signed the Michigan Wilderness Heritage Act into law, establishing federal Wilderness protection for 91,535 acres of Michigan's National Forest lands.

I wrote a summary of the journey for Mackinac Chapter's *Mackinac* Newsletter in February 1988. It closes with this paragraph of appreciation for the many volunteers who worked to pass the bill.

> Together, we have done no small thing. As long as our democracy persists, these fragments of wilderness will remain wild. The loons will call at Sylvania, Lake Michigan will wash against an unspoiled shore at the foot of Nordhouse Dunes, and every spring, the waters will thunder over the falls on the Yellow Dog at McCormick. This is our wilderness to enjoy, to cherish and to learn from, and it will be our children's and their children's. Thank you all for making a dream come true for all of us.[6]

CHAPTER SIX

Who Owns the Sacred and Beautiful?

Beyond Michigan Wilderness, I worked on many other public land issues, including those related to the relatively new national lakeshores. Congress had established four Great Lakes National Lakeshores as units of the National Park System in the 1960s and 1970s, and in the 1980s, their growing pains surfaced on our agenda at the Sierra Club Midwest office. The lakeshores were the product of a lengthy process, dating back to the history of the national parks themselves.

The United States established national parks in places where people had known these places deeply—some for centuries, some for decades—in one context or another. The complex history by which parks became part of the American idea left us with unresolved issues that we wrangled with as we tried to protect the gems of the Great Lakes. Whose lands are these? What does public land mean in our culture? Whose heritage do we respect and celebrate? How did anyone come to "own" the sacred sands and places of haunting beauty in the first place? Who has a right to be here? Who has the right to decide?

The grid imposed on lands northwest of the Ohio River by government surveyors in the early years of the republic had little meaning in terms of watershed, Indigenous homeland, or any other ecological or social context, but it looked logical on a map. Lands within these proclaimed territories became real estate, a framework that would shape settlement and appropriation of Indigenous homelands.

As the new nation matured in the late nineteenth century, settlers reshaped the land itself. They transformed forests, savannas, and prairies into farms, towns, and early industrial sites. The rapidly changing landscape awakened public interest in preserving something of the presettlement undeveloped character of the land and its natural beauty. Whether driven by the romanticism and the frontier nostalgia that Cronon described, or the entrepreneurial anticipation of tourism dollars promoted by the railroads' call to "See America First!" the idea of conserving scenic wonders and natural beauty took hold. The railroads led the successful effort to establish Yellowstone as "a park and pleasuring ground" in 1872. No one asked permission from the many Indigenous people of the region when proponents drafted the boundaries for the legislation. The federal government had already claimed this as American soil. One account on the early history of Yellowstone simply described the Native peoples as having been "evacuated" from the park.

In California, John Muir rose to make the case for national protection of Yosemite, under siege by sheep ranchers, loggers, tourism hustlers, and hydraulic miners. It became the second national park in 1890. This effort also laid the framework for growing public interest in land conservation through the Sierra Club and other groups. Other flagship parks followed. The parks were not only a response to preserve and protect natural wonders, but an experiment with the relatively new American concept of public ownership of land. These lands were not what my Canadian friends call the "Crown lands," but the *people's*. The question of *which* people under what circumstances is as fraught as any other in the long and complex history of rights and freedoms in American society.

As Congress established more parks, their disparate mandates, management, and funding were problematic. In response, Congress established the National Park Service (NPS) in 1916—a new bureau within the Department of the Interior to manage all national parks, monuments, and historic sites. Public enthusiasm for the national park idea drove the system's expansion, and Congress established the first eastern unit—Maine's Acadia National Park—in 1919. In 1933, President Hoover authorized Congress to find a candidate for national park status "to conserve a prime example of North Woods Wilderness," which drew attention to Isle Royale in western Lake Superior. President Franklin D. Roosevelt's signature established Isle Royale as an official national park in 1940—the first in the Great Lakes region, and for a long time, the only one.

During the New Deal years, NPS surveyed U.S. coastlines looking for suitable areas for new national parks. This produced only one new park: Cape Hatteras National Seashore. Then, in 1957, a grant from the Mellon Foundation supported a new survey of the Great Lakes and Pacific coasts. The survey report, *Remaining Shoreline Opportunities*, identified several promising areas in the Great Lakes region. Of the scenic and natural areas reviewed in this extensive study, only four made it through the designation process to become national lakeshores.

The first, in 1966, was Pictured National Lakeshore—a linear landscape of colorful sandstone cliffs, soaring sand dunes, and birch-lined beaches on Lake Superior's southern shore, stretching from Munising to Grand Marais. Later that same year, Congress finally protected the golden dunes, beaches, swales, and bogs along Lake Michigan's southern shore by establishing the Indiana Dunes National Lakeshore.

Then, in 1970, Congress established Sleeping Bear Dunes National Lakeshore, a landscape of towering dunes, crescent beaches, lush forests, sparkling inland lakes and streams in lower Michigan's "little finger" region. Also, that year, Congress added another jewel to the system: the Apostle Islands National Lakeshore, a forest-green archipelago of twenty-one sandstone islands in Lake Superior's Chequamegon Bay accented with historic lighthouses, rocky cliffs, and sea caves, surrounded by the clear, cold, waters of the big lake.

The designation battles were all heroic, and others can tell those stories better than I. I entered the picture when these young and controversial lakeshores were all grappling with public disputes about land acquisition and park management. Already in love with Sleeping Bear Dunes and Pictured Rocks, when I took on regional responsibilities, I soon discovered the further delights of the Indiana Dunes and Apostle Islands. Behind the scenes, these lovely lakeshores faced a tangle of administrative and appropriation hurdles. Because the Wisconsin delegation wasn't part of my regular portfolio, I didn't have assignments for Apostle Islands; the other three lakeshores offered plenty of challenges.

The western parks started out on what the federal government classified as public land. The lakeshores, and many of their saltwater cousins, the national seashores, had been cobbled together out of a patchwork of state or local parks and considerable tracts of private land. The prospective lakeshores were dotted with homes and cottages as well as large acreage from farms or private timber holdings, and even parcels of industrial lands in the case of Indiana Dunes.

Carving out the national lakeshores from private land in the twentieth century carried all the historic baggage of conquest. It also faced a parallel set of ecological hurdles resulting from more than a century of settlement and the rise of the industrial age. Many of the jewels of the lakes were also the guarded haunts of the wealthy, who summered along breezy shorelines in Victorian cottages dating back to the days of ferry and train travel. Other landscapes, such as much of Lake Michigan's south shore, had been purchased by growing industries—especially steel—for future expansion.

After World War II, the automobile and the nation's expanding road system made these once-remote destinations accessible to a wider cross section of the public. As more people sought the beaches, cool northern waters, and fresh air of shores and forests, federal and state agencies responded by creating more parks and recreation areas closer to the nation's population centers.

Park planners had drawn boundaries for the lakeshores and seashores to achieve a delicate balance of management cohesion across ecological goals, established land uses, and local political acceptance. Every private parcel within a future park meant a potential minefield of legal hurdles and public conflict to bring it into eventual public ownership, and every purchase would require adequate federal funding. Nevertheless, we attempted, in the modern era, to piece together fragments into whole landscapes again, worthy of national park system status.

New parks require management plans, and as writers know, the first draft can be the hardest. These plans were open for public scrutiny, and NPS solicited input from written comments and public meetings. For those who didn't want the federal presence in the first place, those early management plans were an opportunity to revisit sore spots and rally unhappy public sentiment. In fairness, park proponents were also actively engaged in the public planning process, and sometimes active critics of park plans.

At one point in the early debate about Sleeping Bear Dunes, a Michigan paper ran an editorial under the headline "For the Few, or For the Many?," getting to the heart of one of the thorniest issues in terms of converting natural treasures to public lands. (The editors argued for "the many.")

In the late 1970s, I attended a large public meeting on the Sleeping Bear Dunes management plan held on MSU's East Lansing campus. As I listened to the long parade of testimony, which included fervent arguments for protecting the area from overuse, I was thunderstruck by the comments of one speaker. His general argument was that access to enjoy places such as Sleeping Bear Dunes and the adjacent Glen Lakes was a privilege that we must earn, and that opening these lands to the public masses would degrade the experience for those who had earned their place as landowners in this gorgeous corner of Michigan.

I can understand the anxiety about hordes of people tromping over the places you treasure. We all want to protect the places we love, and the "love-it-to-death" effect of too many visitors in fragile places was already evident in places like Yosemite and Yellowstone, which were meant to be shining examples of land protection. Still, during his remarks, my inner "House of Commons" began to rise in protest. It seemed to me that he was saying you either had to be born into the privilege of enjoying these exquisite hills and shores, or accumulate the wherewithal to "buy" your way in. The deserving, apparently, already had a slice of the scenic pie, and didn't want us to mess with that arrangement.

I admired his courage for making a public statement defending his views but was disturbed by his assertions, as were many in the packed auditorium. I didn't envy the Park Service employees who had to weigh all this input in their planning process. It wasn't easy at Sleeping Bear Dunes; it wasn't easy for any of the national lakeshores.

Management plans cover many facets of park use and operations. For example, an early controversy at several parks was the number, location, and scale of campgrounds, and whether they would have electrical or other services. Plans also delineate things like priorities for restoration of degraded landscapes, historic preservation, and facilities development, as well as staffing, interpretive services, and much more. I'll come back to the plans in a bit, but part of what drove public controversy was that park managers were crafting these plans while also trying to accommodate visitors and public expectations—"inventing the plane while flying it."

One of the hardest parts about flying that incomplete plane was battles over land acquisition for the private parcels in these young parks. Like hot magma under a thin crust—there was no safe place to set foot in these debates without the likelihood of getting burned.

Political considerations to accommodate a wide range of interests had influenced which lands were included or excluded within park boundaries. One example is the buffer zone for Pictured Rocks. The zone is a strip of privately owned land between the park and the rest of the world that doesn't allow development such as roads or commercial buildings but does allow commercial logging. It floats in something of a no-man's-land between the protected status of the park and commercial use. What exactly are you supposed to do with a buffer zone?

Indiana Dunes faced the largest challenge with historic and ongoing land use. As early as 1906, the steel and other industries were converting big swaths of the Indiana shoreline from dunes and woods to industrial sites. Try to picture the sprawling otherworldly steelmaking contraptions in places that are now Gary, Indiana, and its neighboring communities, as sites for popular and prestigious hunting and gun clubs in the verdant woods along the Indiana shore.[1] The industrial developers took down and carted away the dunes to build Gary. But the national park idea wasn't far behind. In 1908, architect Jens Jensen, ecologist Henry Cowles, and others in Chicago's Prairie Club first floated proposals for "a great Midwestern park."

Prominent Chicago businessman Stephen Mather was also a Prairie Club member and a national park advocate. Familiar with Chicago's nearby natural assets, Mather proposed Sand Dunes National Park to stretch from Gary to Michigan City, Indiana, and held the first hearings in late 1916.[2] The following year, Woodrow Wilson appointed him to serve as the first director of the NPS. Public and even industrial momentum began to build for the park idea. Then World War I shifted congressional priorities. In the 1920s, a state park was proposed. Next, the nation grappled with the Depression, followed by World War II and the rising demand for steel and industrial capacity. These all shifted the park idea to the back burner.

Meanwhile, anticipating the need for future expansion, various companies were purchasing large tracts of land on and near Lake Michigan after the war. Realizing that that industry had plans to expand eastward from Gary into the remaining undeveloped portions of the Indiana shoreline, Dorothy Buell organized the Save the Dunes Council in 1953. The council revived the idea of a national park, working with Illinois senator Paul Douglass to pursue the idea in Congress.

One of the biggest hurdles to a national park designation was Indiana's dream of an industrial port on Lake Michigan. Since the Northwest Ordinance, business and government leaders viewed a transportation link between the Great Lakes and Mississippi watersheds as economically important. Indiana wanted to be that link, with an industrial port on Lake Michigan. Plans for a national lakeshore didn't mesh easily with the idea of a major industrial port. Although Michigan City had a small port, the site wasn't considered ideal for a major hub. Even so, the State of Indiana really wanted a Lake Michigan port, Bethlehem Steel wanted a facility on the shoreline, and National Steel wanted increased shipping capacity. In the early 1960s, Interior Secretary Stewart Udall helped forge what became known as the "Port/Park compromise." Thus, the park's 1966 authorizing legislation carved out the industrial lands and port in the center of the Indiana Dunes National Lakeshore. The result, according to my friend Lee Botts—a passionate champion of Indiana Dunes—was "two huge steel mills and $400 million of federal money to subsidize the port."[3] Lee lamented that the some of the most important lands in the envisioned Sand Dunes National Park were where the port and the two adjacent mills now sit. These lands included the sites where the pioneering ecologist Henry Cowles conducted studies of natural succession in the 1890s and which some described as the "birthplace of ecology." This area, known as "Dunes Park," was only thirty-five miles from Chicago and had been virtually undeveloped prior to the push for the port. After the compromise, it became a five-mile industrial strip flanked, eventually, by two halves of what could have been a whole national park. Bethlehem Steel "gave" the dunes on its site, in the form of bulk sand loaded on barges, to Northwestern University, which used it as lake-fill for campus expansion.

To date, Indiana Dunes is the only instance where Congress authorized a new national park wrapped around five miles of heavy industry. As Lee said, with the memories of one who tasted the bitterness of the loss, "For twenty years, I could not drive down highway 12. That should have been *the heart* of the national park."[4]

In addition to the big challenges to landscape integrity, these parks faced countless smaller ones that became large headaches. As a means to create a park out of patchwork of public and private land, the NPS had a handful of options to acquire the private lands, called "inholdings," within park boundaries. Many inholdings were private residences or farmlands. The terms of acquisition would be bandied about in meetings, hearings, and news stories for decades. The simplest acquisition option was "willing seller/willing buyer," which meant someone was willing to sell their land to the NPS for a mutually agreed upon price, and the NPS was willing to buy it and had the funds to do so (which was often the challenge). This was, for all parties, the simplest option in terms of the transaction, but not necessarily in terms of reaching agreement on property values and aligning that with available acquisition funds. If your neighbors got a better deal on their property, it was fodder for community friction and resentment.

Another option was the "leaseback." Many property owners didn't want to leave their places in these well-loved landscapes, and who could blame them? This option meant you sold your land to the Park Service, and it leased the land back to you as a tenant—sometimes for as long as twenty-five years, sometimes longer. For those who had no interest in selling or leasing, the other options included holding out to see what kind of deal you could get, litigation, or eventually facing condemnation. Under condemnation, the federal government could use its power of eminent domain to seize private property for public use—the choice of last resort in terms of community relations.

A perpetual challenge the Park Service faced was the ups and downs of federal budgets and congressional appropriations, making it nearly impossible for them to manage the land acquisition challenge with a consistent and equitable process. The agency couldn't count on needed funds from one Congress to the next. A delicate negotiation might fall apart if the appropriations weren't available when it was time to close the deal, and people tended to blame the Park Service for failures, not Congress or the president's budget recommendations. Funds for dismantling and removing structures were also often inadequate.

No lakeshore was immune from the conflicts surrounding land acquisition. The anger, mistrust, and fear of local property owners affected by the parks ran deep, and land acquisition affected hundreds of property owners across the lakeshores. At one point on the Sleeping Bear journey, we were working to protect portions of the forested hills that rise above the north shore of the Glen Lakes for ecological and scenic values. The park boundary arcs around the western shores of Little Glen Lake, but not the eastern bay—Big Glen Lake. Purchasing likely building sites on the hills from willing sellers was one of the options under discussion. Some in the community swore this was part of a plot to encircle the Glen Lakes with a network of park holdings and thus create conditions enabling the NPS to declare the homes on these lakes inholdings that would then be eligible for condemnation. There wasn't a scrap of truth in the "encirclement" theory, but paranoia was high. When relations are this highly charged, it is a poor basis from which to build lasting community relations, or to make wise decisions for the land.

My friend Lee had a leaseback home in the Indiana Dunes that an arsonist destroyed in 1981, but she bargained for and secured the lease on another available leaseback property, which became a new hub of family and environmental activity in 1983. When the lease with the Park Service ended twenty years later, that house and neighboring buildings were torn down. In a commentary in the *Chicago Tribune*, Lee's daughter Beth wrote one of the most moving pieces I've ever seen about what it meant to let go of their claim to this place:

> No one could have loved a patch of ground more than we loved that stretch of dunes and marsh. No one could have cherished any sight more than my mother cherished watching from her window while the lake changed its grays, greens, and blues, turned red under sunsets, rose in storms, piled up white ice, lay serene under a summer day.
>
> Losing a cherished place has its pangs, for sure. But peaceful surrender has its joys . . . the old place had become part of a slightly more continuous grandness, a natural grandness that existed long before anyone built a house there and that, thanks to the national park, will be there long after all of us are gone.[5]

The Botts family and the hundreds of other families who let go of pieces of the shore to make that grander whole possible are gentle heroes whose gifts of grace will likely never be fully appreciated by future generations who walk the same trails and play on the same beaches. Their legacy is that untold thousands may come to love these places, too, and love them enough to want to pass them on whole to their own descendants.

Public land and public rights in a private-property culture are tricky things. In my reading, property is not one of the unalienable rights that Thomas Jefferson asserted that we are endowed with by our creator (although Jeffersonian scholars might argue it was implied). It is not part of the "life, liberty and pursuit of happiness" trio, but you'd think it was from the passion our culture invests in it. When you wrestle with it, who really *owns* lands taken by multiple conquests, land grabs, and opportunistic settlements over multiple centuries? Even if there were an equitable way to resolve this, I'm not sure it's the key question.

Ownership doesn't necessarily resolve the question of how we live well and respectfully in relation to a place and each other. That's the genuinely hard work of a diverse democracy living in a global context.

Time and again, I return to a preference for the long-term common good over near-term private (and perhaps also public) interest. I retain an abiding respect for the rights and needs of individuals, local communities, and the people whose land this was long before my ancestors got off a boat. Sometimes it feels like we can't accommodate all those needs, but it doesn't mean we can't try very hard to do so. Here, if we must make difficult choices, I side with Aldo Leopold's criteria, in which he argues, "A thing is right when it tends to preserve the integrity, stability and beauty of the biotic community. It is wrong when it tends otherwise." When we tend the garden of life, we all benefit.

CHAPTER SEVEN

Preservation, Enjoyment, and Managing Expectations

Let's return to those management plans. Beyond land acquisition, lakeshore managers faced the double-edged sword of the agency's dual mission: "The National Park Service preserves unimpaired the natural and cultural resources and values of the National Park System for the enjoyment, education, and inspiration of this and future generations."[1]

It is hard to preserve an area and leave it unimpaired while you are trying to provide enjoyment for lots of people, and this paradox has dogged the Park Service for a century. Tourist accommodations place one set of demands on a place; protecting rare plants, wildlife, fragile natural features, and historical sites place another. The parks are supposed to achieve both ends, at the same time in the same landscape.

During the 1984 congressional wilderness tour, we wrapped up our multiday marathon at Pictured Rocks National Lakeshore, where the Park Service had just finished building its new restroom facility near Miner's Castle. This building had been controversial, given its fort-like architecture and the size of its footprint on the site. One disgruntled ranger privately bemoaned it as the "monument to defecation."

Regardless, one of the Senate aides on the tour was thrilled to see a real bathroom with flush toilets and running water only a short walk from the paved parking lot. We had traveled to a half-dozen remote locations via bumpy gravel roads and urged her along on walks through mosquito-dense forests with barely a trail. As a polished city gal, she had managed all this with considerable grit and grace. When we reconvened at our van following the rest stop, she beamed, "Now I get the difference between parks and wilderness. Parks have sidewalks and bathrooms!"

The sidewalk and bathroom factor, as well as the restaurant, visitor center, hotel, gas station, and paved-road conveniences, put pressures on parks that aren't part of the Wilderness management portfolio. It also raises expectations that these amenities are what a proper national park provides. Many local communities felt that park promoters and the federal government had sold them a bill of goods about the tourism benefits of the lakeshores when the amenities were slow in coming, and the public didn't beat a path to their door after initial designation.

While appropriations were frequently inadequate for land acquisition, as noted, they were also inadequate for facilities (that monumental restroom opened eighteen years after the park) and other park functions. I remember a 1989 walk at Sleeping Bear Dunes where the park superintendent was excited to show us a new native dune plant restoration area—made possible because they had finally landed funds for this long-delayed project. Also at Sleeping Bear Dunes, there was once a plan for a living history village in Glen Haven, dashed somewhere along the way by the budget cutters' red pens.

Caught between the elevated expectations of the original legislation and early management plans, and the realities of the federal appropriations process, it sometimes felt like Park Service staff just couldn't win. Still, people who believed in protecting the lands for the future and public enjoyment tried to move forward in constructive ways.

The hostility and frustration around land acquisition and management plans at the various lakeshores was a source of considerable anguish to local communities, the Park Service, and to those who just loved these splendid places and wanted to see them remain unharmed. Supporters and park staff often paid dearly. One park superintendent told me stories of how the local kids routinely harassed his children in school. Other park staff spoke of being vilified in letters to the editor, shouted at in public meetings, or shunned at the grocery store.

The disappointment (and downright resentment) that tourism dollars were not meeting local expectations was certainly a prevalent sentiment around Pictured Rocks for years, where funds for facilities and visitor services were painfully slow in coming. As the lakeshore's own administrative history notes: "Pictured Rocks has shared with the other lakeshores the problem of being born in an era of inflated expectations and coming of age during an era of constrained federal budgets."[2] We also need to remember that Pictured Rocks as a park was conceived of in the "see the U.S.A. in your Chevrolet" days of the 1950s and early 1960s, but was hatched just before the "born to be wild" era on the cusp of the 1970s. Less than a decade after initial plans had been put in place, demand for wild country—especially in the populated East—was overshadowing the original concept of experiencing nature through car windows on a scenic highway.

Early plans for Pictured Rocks included a proposed scenic shoreline road. It was included in the 1966 authorizing legislation, but controversial even then. This road would have been a two-lane, paved undulating strip traversing the forty-two-mile length of the lakeshore with pullouts and outlooks perched on top of the cliffs along the way. It seemed like just the ticket to bring in hordes of tourists and their dollars in the economically depressed U.P. To those who sought the untamed shores of Lake Superior, far from the grind of shifting Winnebago gears, the road was anathema that would carve up and pave miles of backcountry, deface the ancient cliffs, and destroy the wild and rugged character of most of the shoreline. It was a conflict in generations, values, and dreams.

Forty-two miles of construction along eroding sandstone cliffs, shifting sand dunes, and wetlands would have been a very expensive undertaking. While it is one thing for Congress to authorize a project, it is yet another for Congress to appropriate the funds to pay for the project; thus, without funds, the road was only a concept in the late 1970s, and still under debate.

As part of the "no-road" crowd, I have vivid recollections of people yelling at me, flinging insults, and sometimes even threats, especially at meetings in or near Munising. People had a lot of anger to vent about what they believed was a broken promise of prosperity by early park proponents and the NPS. In that context, many considered road opponents to be part of the problem and exacerbating the broken promises. I appreciate that local residents had a right to their viewpoint and understood how they concluded that they'd been snookered. Still, for the longest time, I dreaded being in Alger County for fear of more vitriolic confrontation.

As the lakeshore became better known and popular, its constituency of car campers, boat tourists, backpackers, and anglers grew to love it as it was, which was largely undeveloped except for the fading footprint of historic logging and isolated former home and cabin sites. As this constituency grew, so did demands to scrap plans for a shoreline road. In the late 1970s, we had persuaded Congressman Carr of Lansing to introduce legislation to deauthorize the shoreline road. Eighteen years later, a version of that legislation finally passed. If we did nothing else in our careers beyond helping to prevent bulldozers from gouging a two-lane corridor through the maple and birch forests along the sacred shores of Pictured Rocks, Bob Carr and I and the stalwart others who carried the torch will have done enough.

The eventual compromise was a paved version of old H-58, an existing inland road corridor. It serves those who want to zip through the lakeshore in their cars, while providing access to the campgrounds, trailheads, and spectacular overlooks and beaches between Munising and Grand Marais. The lakeshore trail that weaves along parts of the sandstone escarpment, nearby forests, and the dunes, remains for those on foot, connecting with other trails within the park boundaries, and serving as forty-three miles of the North Country Trail that reaches from Vermont to North Dakota.

The road compromise seems to have worked, saving vehicle-travelers from the automotive version of sand traps on the previous version of H-58. It certainly hasn't dampened visitor numbers. In recent decades, kayakers have swelled the ranks of visitors. Guided tours enable visitors to reach the isolated beaches and sea arches in lake-worthy sea kayaks. (Please don't attempt this in a recreational kayak—it's just not safe on Lake Superior.) And for the many people who lack the interest or ability to hike, camp, or kayak, and prefer the short walks from their cars to the outlooks for the view and some photos, this also makes that experience available.

Still, the love-it-to-death threat has taken its toll on the national lakeshores, as many feared it might. One of my favorite professors, Paul Risk, once served as a park ranger at Grand Canyon National Park. One day in class, he told the story of failing to restrain his attitude when a visitor jumped out of their car and asked him, "How long does it take to do the park?" His response was, "It depends on what you plan to do to it." Too many people do too many unfortunate things to our beloved parks.

One day in the late 1980s, I received a call from Larry Lemanski, a passionate defender of Michigan wild country who shared my love for "the Rocks." He was practically in tears. He described the erosion and damage to the sandstone bluffs at the Mosquito Beach and Chapel Beach campgrounds from foot traffic and the curious hands and careless feet that broke off portions of the brittle sandstone shelves. He talked about the compacted forest floor, trees stripped of branches for firewood, and scars on the ground from fire rings. "People are destroying the place," he urged, "We've got to do something!"

The Park Service, under their split directive, has tried. They moved campsites inland from the fragile shoreline, banned campfires in the backcountry at sensitive sites like Mosquito and Chapel Beach campgrounds, and regulated the number of backpacker permits. Still, it only takes a few jerks, innocent or intentional, to damage fragile places, and a Park Service staff plagued by grinding budget cuts can only cover so much ground.

When Pictured Rocks celebrated its fortieth anniversary in 2006, the Park Service reported that 450,000 people visited this wondrous and fragile place in that anniversary year, generating an estimated $19 million a year in economic benefits. This seems like a respectable shot in the arm for the local economy from a park some lamented would only benefit a few hippie backpackers. The numbers continue to climb. In 2019, that number had doubled to an estimated $38.1 million in local gateway communities as reported by the *Mining Journal*.[3] These expenditures supported 477 jobs and $11.7 million in labor income, according to that year's NPS Visitor Spending Effects report.

When I made a brief visit to the Lakeshore in September 2020, the size of the crowds at one of the trailheads shocked me. The parking lot was overflowing, and expectant hikers had crammed vehicles on the side of the road for at least a half mile above the lot. We almost turned around and left, but we were trying to heal from a family tragedy, and I felt a desperate need to sit with the waves and sandstone at Mosquito Beach. All the visitors were likely seeking time in nature's beauty during these tough months of the pandemic, but the crowd felt like a mall at Christmas. We took a trail less traveled, and left most of the crowd behind, but it left me worried about the impact of all those feet and cars, including mine, day after day. The lakeshore set a record of 1.2 million visitors in 2020. Larry would have been aghast.

The Park Service now faces the challenge of how to manage the crush of visitors in these small and ecologically fragile places without the gated entries of a park like Yellowstone. I understand entry fees are under discussion, and campgrounds require advanced reservations many months ahead if you have any hope of camping within the lakeshore boundaries. I only hope these hundreds of thousands of visitors will love Pictured Rocks and other lakeshores gently. Somewhere, Larry Lemanski's spirit is out there keeping an eye on how people are behaving when they are in the lakeshore, and I hope they will honor his passion for the place by respecting the integrity of its wholeness.

Over time, something remarkable also began to happen in the larger landscape. Nature was knitting together the living fragments of the patchwork pieces into a healthier whole with a helping hand from the Park Service and countless volunteers. Restored dune grasses helped stabilize fragile slopes, reducing erosion and providing habitat. Maturing forests began to support important species like warblers. Native marsh plants and wildflowers returned where staff and volunteer crews painstakingly dug out invasive species. The lakeshores were and are becoming more vibrant, diverse, and resilient ecological landscapes again, so let's not inadvertently trample them with our enthusiasm to be within them. Similarly, as local communities have gained from the economic value of the parks, the flames of controversy surrounding the property issues have died down bit by bit as time and generations have passed.

These national treasures—global treasures—belong to all of the earth's inhabitants (and not just the humans). In 2001, I watched a focus group about the Great Lakes in which a group of women discussed a variety of issues. The researcher asked, "Who owns the Great Lakes?" There was a pause, and then one woman slowly raised her hand, pointed to the ceiling, and said, "God does." Other women nodded their heads, and the conversation turned to the notion that the lakes are gifts held in common for the benefit of all, not commodities to be bought or sold. I would argue that the scenic and natural treasures on the shores deserve that same status. Public land is a mechanism—not perfect, but one of the best tools we have—by which we can safeguard, to the best of our abilities in a democracy, places that provide a common benefit to all.

In addition to the battle of the Pictured Rocks road, I worked on dozens of other national lakeshore projects. We fought to secure funding for land acquisition and operating budgets and to support land swaps and additions that would improve the integrity of the parks. We read stacks of proposed management plans and submitted detailed comments and participated in dozens of meetings on issues ranging from campground development to wilderness studies to visitor facilities. My colleagues and I met with park and congressional staff to find ways forward, and every now and then we walked the dunes, forests, and cliffs with Park Service superintendents and rangers, congressional staff, and others who were in roles to decide the fate of these precious lands.

The lakeshores were their own best advocates. When we were working on plans to protect wetlands in the Bow Lakes Basin in Sleeping Bear Dunes, the Park Service offered a site visit, and thus an opportunity for another trip in which we'd drag the congressional aide through the woods. As we set out on the trail, the clouds gathered and rumbled, and the sky opened. It drenched our group, which included park staff, Senator Riegle's aide, and me as we trekked in to look at one of the ancient bogs firsthand. Whatever rain gear we had was inadequate. We were soaked to the bone but made it to our destination, where we engaged in a soggy dialogue about the value of the place, the need for authorization and funding, and all the technical stuff of saving a piece of land with public money.

We hiked back out, all conversation lost to the drum of the rain. The Park Service staff dropped us off at our respective rental cars, and they headed home to shelter and dry clothes. The Senate aide and I pondered what to do about our sopping selves before climbing into our cars and heading to other destinations. I kept thinking, "He's such a decent guy, and we've dragged him out here and made him slog on slippery trails to a bog you could barely see in the storm. Now we're soaking wet and shivering out here in a parking lot. He's going to remember this as the lousy experience in the Dunes, and there goes this proposal." But I'd underestimated him and the bog. As we stood dripping under the eaves of a building adjacent to the parking lot, the torrents still coming down, he said, "You know, that area hasn't been altered since the glaciers were here."

"Yes," I replied. "I guess that's right."

"Isn't that incredible?" he responded. And I knew the place had won another friend. Today the area is a protected part of the Lakeshore.

On rare occasions, I got to be a visitor or camper myself, but when you carry around a "to-do" list in your head every time you set foot in a park or Wilderness area, it can defeat the goal of getting away from it all. When I left my field rep role in 1991 after more than twelve years in the thick of federal battles, I didn't know how much I needed to move on from the intensity of conflict and the roller coaster of hopes and plans that rose and fell with the whims of changing administrations and Congress.

Just being in the lakeshores was a reminder of all the conflict and all that remained to be done, and it was emotionally taxing. I needed some distance from the battlefield, from the wounds, disappointments, and outright failures, because for some reason the victories never erase the losses, and we had our share of both. It was time to let go for a while. I knew there would be others who would take up the cause of caring for these places, and they have.

Many decades after the bold move to establish the national lakeshores, the lands within them are still largely intact, albeit over-trodden in some places. Where planners scrapped the idea of a national lakeshore early on, such as in Wisconsin's Door County, we have a comparative landscape that tells the story of what happened to an area without national protection. Portions of this finger of land, which reaches northward between the Bay of Green Bay and Lake Michigan, are still gems of Great Lakes natural beauty—especially its two state parks. But the march of development has transformed so many once-empty shorelines, orchards, and farms into developed real estate.

The contrast between Door County and its sister shoreline across the lake demonstrates the power of the National Park System to protect places of great beauty and historical significance from what a Door County real estate road sign promoted as "40 acres, great development opportunity!" I know the birthing of the national lakeshores was painful, but every time I visit Door County, I believe it was the right choice.

Another area that made the finalist recommendations in the original *Remaining Shoreline Opportunities* report represents a different alternative. The vast Huron Mountain Club, considered the premier opportunity in that study, remains a gated and patrolled private enclave embracing twenty-two thousand acres of inland forests and miles of Lake Superior shoreline that is "invitation only."

In August 2021, the *New York Times* printed a guest editorial: "Give the People What They Clearly Need: More National Parks."[4] It was a response to the record-setting crowds (327 million visitors in 2019) *before* the pandemic as well as the crush during. It recommended redesignating some national monuments as national parks to spread the excessive love around to some of these spectacular but less known national treasures. OK, but that would still place pressure on an already-strained federal agency to respond to the expanding national passion for outdoor experiences. Encouraging people to explore the nation's National Forest Recreation Areas and National Wildlife Refuges with visitor accommodations (as well as state parks and forests) would also reduce pressure. But we've done such a successful job of "branding" the national park experience that the big parks are often first on the list of outdoor destinations. It has worked so well that people "collect" park visits on life lists in passport books for bragging rights. The crowds are part of the backlash of success.

That success hasn't come easily. Beyond inadequate resources, the lakeshore challenges were about conflicts in underlying values and making difficult choices between accommodating the needs of recreational visitors, fragile ecosystems, and the rights of people with ancient and recent connections to a place. Not every acre can be all things to all people all the time. The future of parks depends on restoring not only physical landscapes but whole human communities in synergy with the wild fabric of nature. These are things that lines on a map can only begin to address.

Like thousands of others, I love the sweeping sands and blue water vistas of Sleeping Bear Dunes. I have spent golden summer days on the stretching beaches of Indiana Dunes with a happy toddler playing tag with the waves. I have slept under starlight in the exquisite isolation of Julian Bay in the heart of the Apostle Islands, and I am connected in a way that transcends love to the sandstone shores of Pictured Rocks.

I honor and respect the spirit of those who walked these same shores in ancient times, and their living descendants, whose stories are older and deeper than mine, and whose connection I can only approximate with that of one brief lifetime. I thank the people who struggled long for the legislation to establish these lakeshores, and those who first took a leap of faith that it was possible. Their actions have made my cherished experiences, and those of millions of others, possible. I caution that a simple act of Congress never fully protects a place. Vigilant stewardship (and public resources) in perpetuity is part of the bargain.

The gift of the national lakeshores was that when I was ready to return to them again many years later, they were still there, offering a welcoming embrace of beauty, shining on the shores of the Great Lakes. And this same embrace is open to the many, not just the few, to celebrate life, experience freedom, and pursue a little happiness in beloved places we hold in common.

PART THREE

To Protect the Water

If there is magic on this planet, it is contained in water.

—Loren Eiseley

CHAPTER EIGHT

Nukes, Dunes, and the Lady of the Lakes

When I dug into Great Lakes issues, I quickly learned that my generation stood on the shoulders of pioneers in Great Lakes activism whose work since the 1950s had shaped ours and still shapes the efforts of those who follow. Many of those unsung heroes were women who served as role models for young environmentalists like me who rose during the height of feminism. In politics, they tell you to dance with those who "brung you," and these women knew how to dance, work, and succeed. I'm glad for those who brought me with them.

My story would be incomplete without acknowledging the influence and impact of one of those champions in particular: Lee Botts.

Lee was a compact powerhouse of a woman whom I first met in the early 1980s. Born and raised in Oklahoma and Kansas during the dust bowl years, in 1950 Lee ended up in the Hyde Park neighborhood that is home to the University of Chicago, where her husband pursued his master's degree. For a time, she was a copyeditor for the *Journal of the American Medical Association* (*JAMA*) and then went on to write for and edit the *Hyde Park Herald*, the local weekly newspaper. When editing for *JAMA*, she worked on an article by a scientist named Jonas Salk—it was his paper announcing the polio vaccine. Lee somehow had the fortune or tenacity to find herself right in the middle of auspicious moments where worlds are about to change.

In Chicago, Lake Michigan is smack in your front yard and hard not to notice. In the 1950s and 60s, while she was working *and* raising four children, she got involved with local environmental issues—especially issues related to the Chicago lakefront and the battle to create a national park in the Indiana Dunes. The nearby dunes on Lake Michigan's south shore are a great place to take a houseful of city kids, especially in the summer. The Botts family rented cottages and made friends as they became regulars along the shore.

Opposition was rising against plans by the steel mills and power plants to gobble up more of the shoreline between Gary and Michigan City, and this inspired renewed efforts to create a national park instead. In 1959, Lee joined the Save the Dunes Council to help protect what remained of the threatened dunes. The drawn-out battle finally succeeded. Lee tasted that 1966 victory with the ranks of others who had loved and labored to protect this special place, and many joined with her in creating the Lake Michigan Federation as the first environmental organization devoted to the Great Lakes.

As an old saying in environmental circles goes, "All our victories are temporary; all our defeats are permanent." Just when it seemed as if the dunes were mostly safe, along came a plan by the Port of Indiana to build a nuclear power plant in the middle of the new national lakeshore. Named for Joseph Bailly, a long-ago fur trader, this plant was one of up to twenty-four plants that the original Atomic Energy Commission (AEC) wanted to site on Lake Michigan alone.

Like all power plants that produce electricity from steam-driven turbines, "nukes" need a lot of water to create the steam. The process also produces substantial amounts of heat. Plant designers devised the gigantic, squeezed cylinders known as cooling towers to cool the waste-heat—another big demand on water—because discharging hot water into lakes and streams can kill fish and destroy habitat. Depending on the technology, the cooled water can be recirculated (a "closed-loop" design) or discharged into a local water body. Where better, from an engineer's standpoint, than the shore of a vast and cool Lake Michigan to site a nuke—especially one close to major industrial markets? And where better, from the standpoint of human society and the Great Lakes ecosystem, to invoke the precautionary principle?

In simplest terms, the precautionary principle argues, "If you don't know what the consequences of your action might be, err on the side of caution." Or "It's better to *not* do something unless you are confident that it is relatively harmless." So, if there's a reasonable chance of something going wrong with a nuclear power plant on the shores of the Great Lakes and it might affect the drinking water supply of twenty to forty million people, not to mention what it might do to the fish and wildlife, maybe it isn't the wisest place to put a nuclear plant. But decisions about the fate of nuclear plants didn't happen in isolation. The story of the Bailly plant begins with the larger story of nuclear energy in the Great Lakes overall.

The 1970 National Environmental Policy Act (NEPA) introduced the then-revolutionary concept of identifying risks in advance of approving major federal projects. It required an Environmental Impact Analysis for all major projects funded or permitted by federal agencies. That same year, Lee participated in a meeting in Washington, D.C., that she said was among the first where citizens met in person with heads of federal agencies to discuss environmental policy related to nuclear power. Proponents had asserted that the United States would need to double its electricity capacity every six to seven years, which would mean a proliferation of new power plants and transmission lines. This sparked grassroots opposition, and environmentalists at the meeting argued that such a growth rate was unsustainable; they urged a more cautionary approach for expanding electricity generation.

The rising specter of nuclear power plants on the lakes was gathering momentum in 1970, which coincided with the first year of operations for the young Lake Michigan Federation. The nuclear industry had pursued something of a "if we build them, they will let us run them" approach to expansion. Thus, plants first had to receive a construction license from the AEC, and then, when completed, an operating license. The assumption had to have been that operating licenses were a sure thing. Otherwise why would you put the enormous financial investment into building a plant if you thought you might not get to operate it? In the early 1970s, construction was nearly complete for the first generation of large-scale nuclear power plants, and these plants would soon need operating licenses to power up and come online. The Palisades plant near South Haven, Michigan, would be the first large nuclear plant to seek an operating license in the Great Lakes.

The utilities were holding public information meetings, and at one of them, Lee met Larry Spear, a student who had been working with the organization Business and Professional People for the Public Interest (BPI) in Chicago. Over several meetings, the two became better acquainted and began to share information. At the public meetings, the utilities were providing information about the cooling systems for the new plants and how these systems would protect Lake Michigan from thermal pollution. Heated water, or thermal pollution, posed its own threat to the lakes—especially fish habitat—because many fish species can only survive within narrow, cool temperature ranges.

Some plans talked about releasing heated water at depth, others near the surface. The industry's inconsistent and unproven approaches to solving the thermal pollution challenge began to raise more than just casual questions from nearby residents. The uncertainty was a growing concern for the region's new wave of salmon fishers. In 1966, Michigan's Department of Natural Resources introduced hatchery-raised Pacific Coho salmon and Chinook salmon into Lake Michigan to control Atlantic alewives, fill an ecological gap caused by the crash of the lake trout population, and boost sport fishing.[1] In just a few short years, salmon sport-anglers and the related charter-fishing businesses had formed an increasingly vocal constituency to champion this new recreational fishing experience in the Great Lakes.

Lee and Larry noticed that while each utility presentation claimed to be promoting the *best* design for handing the hot process water, every plant had a different design—so, what *was* the best? Mary

Sinclair, a watchdog of nuclear development in Midland, Michigan, was noticing the same thing.

The AEC had built a few small sixty-megawatt plants in the 1960s, as Lee put it, "to set the stage for this glory that was to be nuclear power." These included the Dresden plant, southwest of Chicago, and the Big Rock Point plant near Charlevoix, Michigan. By 1970, twenty-nine nuclear power plants had already been constructed on or near the shores of the Great Lakes, and that was just the first stage of bigger plans. Concerned Chicago citizens noted that their community was slated to become the city with more nukes in a hundred-mile radius than any other city in the country, adding to the growing public wariness about this new energy source. A nuclear chill was growing among multiple constituencies in the region.

EPA was a new agency and environmental impact statements were still a new idea. The regulatory structure for nuclear power had been evolving from a patchwork of federal policies that lagged behind the leading edge of the industry. The regulatory structures we take for granted today were in their infancy.

Against this backdrop, the Department of the Interior was still conducting something called "Enforcement Conferences" for the nation's major waterways—meetings required by the Federal Water Pollution Control Act (the precursor to the 1972 Clean Water Act).[2] While the Department of the Interior had designed these conferences for state and local officials and water managers to confer on water quality standards and other matters for interstate waters, citizen groups like the Lake Michigan Federation wanted to be at the table. They wanted to be involved in the decision-making process and were becoming increasingly assertive. The department had stepped into the water oversight role under Secretary Stewart Udall's guidance in 1966, but neither the federal nor state agencies had much experience with engaging the emerging environmental constituencies across the country in 1970—the year of the nation's first Earth Day.

Scientists and water agencies had only recently identified the nutrient phosphorus as the culprit for the wretched, algae-choked conditions in Lake Erie, Saginaw Bay, and so many other places in the

Great Lakes. It was a hot topic, and citizens were clamoring for ways to get action. The multiagency task force that hosted these conferences moved the Lake Michigan phosphorus standards meeting from Chicago to Grand Rapids, Michigan. Lee believed they did it to evade public participation. Still, the citizen groups persisted. They wanted public representation in these conferences, and the Chicago-based Lake Michigan Federation representatives met with Sierra Club members in Grand Rapids to discuss strategies. Both groups showed up for the Grand Rapids meeting, but the federal task force balked, claiming that there just wasn't enough room to accommodate them.

Undeterred, the citizens groups reached out to Michigan's Senator Hart, an environmental champion who happened to be in the state. He agreed to come to Grand Rapids as quickly as possible. Recalling this event, Lee said, "We wanted to take over the Lake Michigan enforcement conference on two issues: the phosphorus standard and thermal pollution." The meeting room was part of a larger room, separated by panel dividers. As part of the "takeover," the citizen groups threw back the dividers and announced that the senator was on his way. The room was then open, and, evidently, so was the meeting, which the convenors quickly recessed for lunch.

The task force regrouped and reconvened after lunch and announced its proposed new phosphorus standard, an important first step to control this serious Great Lakes pollutant. After a dramatic morning, the task force opted to address the thermal standard at a subsequent meeting in Milwaukee. A snowstorm dampened attendance, but not Lee's. There, the agency team charged with developing the thermal standard reported that it was unable to do so. Never shy about calling it like she saw it, Lee commented that the Federal Pollution Control Agency was a "paper tiger," which made headlines in Chicago.[3]

All this foment was percolating as the AEC was about to consider the operating license application for the Palisades plant. Senator Hart asked Lee to assist in organizing a field hearing in Grand Rapids in the spring of 1970 because of the mountain of mail he was receiving from sport fishers regarding the threat of thermal pollution to Great Lakes salmon. At the hearing, the big question among the growing community of nuclear plant opponents was "How do we keep Palisades from getting its operating license?" Lee explained that the AEC had a process for public intervention into the licensing process, but no one had yet used this process. She suggested that the Michigan salmon fishers take the first step and petition the AEC for a hearing. Three weeks later, they called her to report they had thirty-five thousand signatures on the petition and asked what came next. She suggested that they ask Senator Hart for his endorsement, and subsequently, the AEC scheduled a hearing at the Kalamazoo public library.

Meanwhile, Lee had been meeting with a group of sympathetic attorneys in a search for an underlying legal theory to sue the federal agency for failing to protect Lake Michigan. One of the attorneys persuaded his firm to provide a pro bono lawyer, Myron (Mike) Cherry, to assist in presenting the case against a Palisades license. The legal team began to galvanize, supported by BPI and the nuclear industry expertise of David Comey, a Cornell research assistant who had helped stop a construction license from going forward for a plant in New York. The group's team for Kalamazoo included three University of Illinois law students, a Sierra Club attorney, Lee Botts, Mary Sinclair, and a local attorney hired with $10,000 provided by a Michigan donor. With these modest resources, they initiated the first citizens' intervention on a pending operating license for nuclear power plants.

The next big step was something called a "pre-hearing conference." Lee, Comey, and the other strategists agreed to keep the issues broad at this conference until they could figure out the specific points for leverage and really hone their theory. As she told this story, I realized that one of the most important strategic lessons I'd ever learned from Lee was the idea of getting your big game plan down. She was skilled at looking for the powerful lever that could drive big shifts, not just incremental victories. She also knew how to use the public stage. As Lee described it, "The conference was attended by the *New York Times*, the news wire services and representatives of every utility in the region that owned or wanted to own a nuclear power plant." She was prepared with newsworthy quotes and comments to amplify the issue in the media.

The environmental team's legal tactics were based on three big questions about nuclear power regulation. First, did the 1970 NEPA requirement for environmental impact statements apply to nuclear plants (even though construction on these plants had started before the law passed)? Second, who was responsible for regulating thermal pollution from these plants, and third, whose job was it to protect overall water quality in relation to nuclear plants?

The group wrestled with how they could demonstrate that the AEC was responsible for thermal pollution, since the AEC had stated that thermal pollution was a more important issue for public and environmental safety than possible releases of radioactivity. At the same time, the AEC asserted that they had no jurisdiction to regulate thermal pollution. Enter the fresh perspective of a young law student, who was reading through the Atomic Energy Act of 1954. He noted that the AEC was responsible for all forms of energy associated with generating electricity with nuclear power, and asked, "Isn't heat a form of energy?" How can you argue with the laws of thermodynamics?

The assertion that the AEC was indeed responsible for thermal energy as well as radiation became the intervenor's first motion. With this motion on the table, the commission wrapped up the proceedings for day one.

Afterward, Consumers Power, the Michigan utility that would be operating the new plant, invited people to come see the Palisades plant firsthand, which happened to be testing its cooling water pumps during the visit. The plant discharged its process water through pipes that released directly into Lake Michigan. Lee and the others went on the tour. She said, "We watched thousands of dead and dying fish being hurled around in the discharge water. We were stunned. That was it!" The groups resolved to challenge these plants on the lakes.

The hearing went forward, but there was no quick resolution. The environmental attorneys sought an injunction in federal court to prevent start-up of the plant, but the court of appeals told them, "You first have to exhaust the administrative process, but then you can come back." Thereafter, Consumers Power sought to lengthen the hearing. Although it slowed things down, the delay also opened the process for additional scrutiny and public involvement. The "hearing" went on for almost another two years, and the AEC eventually approved the Palisades operating license on March 24, 1971. Although Consumers Power originally designed the plant with a "once through" water system without special cooling systems, all the public scrutiny about thermal pollution led to new requirements in the final license. As a result of the intervention, the plant operators installed cooling towers to reduce the discharge water temperature.[4] The groups may have lost the overall licensing battle at Palisades, but the tide had turned for Great Lakes and American nuclear plants.

As Lee recalled, "By this time, we were getting calls from all around Lake Michigan and the country on how to intervene and engage in this process." The Chesapeake Bay Calvert Cliffs decision of 1973—which really put the brakes on the nuclear industry in the United States—was based on the issues developed at Palisades. In this case, the court of appeals ruled that Congress meant to require environmental impact statements for those plants already built and any more to come. This opened the licensing process for much wider public scrutiny and review. Moreover, the new 1972 CWA required regulations for thermal pollution (and many other types of pollution) from a wide range of sources, including power plants.

The Palisades and Calvert Cliffs challenges shifted the dynamic for all subsequent U.S. licensing processes. They established the precedent of an effective legal intervention by informed citizens and opened the door for intense public scrutiny of all further licensing processes. The Lake Michigan Federation and BPI facilitated interventions by local organizations at other nuclear plants constructed in this era. All eventually received their operating license. In every instance, David Comey found ways to force stricter safety measures. Volunteer scientists also helped inform regulators how to protect fish and other water organisms from being pulled into water intake systems or affected by cooling discharge systems.

Back in Indiana, the Bailly plant was still at the groundbreaking stage. But there were special worries at this location. Under AEC regulations, the owner would have to be able to evacuate about two thousand acres around the plant in the event of an accident, which would include national park land. What if other nuclear plants were located near public lands to gain more control over evacuation? There was also concern about how the plant's construction would affect critical wetlands and other park resources. The Department of the Interior opposed construction. Then the Steelworkers Union joined the Bailly Alliance (which opposed the plant) when they learned that Bethlehem Steel would need to identify "a suicide squad" of workers to stay behind to shut down its nearby steel plant in the event of a nuclear accident. Public opposition succeeded in shelving the plans for a nuclear plant at the Bailly site.

David Comey and Mike Cherry continued to challenge nuclear power at other locations including near Lake Huron, working with Mary Sinclair. They became nationally known to the industry. In 1974 Comey received an EPA award for "services that immeasurably improved the design and safety of nuclear reactors." It was Comey who arranged for the *New York Times* to interview Karen Silkwood about her experience with dangers in the nuclear industry in 1974. Silkwood was a chemical technician making plutonium pellets in the Kerr-McGee Cimarron Fuel Fabrication Site in Oklahoma. She belonged to the Oil, Chemical and Atomic Workers Union, and the union was concerned about risks to workers and quality control of fuel rods produced at Kerr-McGee. She had testified before the AEC about these concerns. In early November 1974, over three consecutive days, tests found that her body contained elevated levels of plutonium contamination.

The interview that Comey had arranged on November 13, 1974, never took place because she died the night before when her car crashed into a culvert on her way home from a union meeting. Many speculated about whether her death was accidental. In a tragic echo, on a weekend trip to Wisconsin in early 1979, Comey died in a truck collision on I-94.

The slow defeat of the Bailly nuclear power plant culminated when the Northern Indiana Public Service Co. (NIPSCO) canceled it in 1981. It was a partial victory. Although dunes advocates had hoped to relocate the existing coal-fired plant away from the shoreline, it stayed put and expanded its capacity. Still, no U.S. utilities have ordered nuclear plants for the shores of the Great Lakes since.

With the long-planning horizon for nuclear plants, most first-generation nuclear plants were completed between the 1970s and early 1990s.[5] According to the U.S. Energy Information Administration, nuclear expansion slowed due to less electrical demand and, following the 1979 Three Mile Island accident, rising capital and construction costs, and public opposition. In the decade that followed, sixty-seven planned builds were canceled.[6]

The handful of new-generation plants built in recent years have far more sophisticated engineering than those 1970s designs, but still face remarkably high capital costs. Georgia Power's new Vogtle plant is a case in point where construction costs have nearly doubled—to more than $28 billion.[7] Today, the nuclear power issue in the Great Lakes region has shifted from new construction to when and how to decommission the existing plants that are reaching the end of their planned lifetime. Decommissioning raises the uncomfortable question, for which there has been no national solution since engineers imagined the first plants decades ago: "What the heck should we do with the radioactive waste?"

Nuclear power's role in addressing climate change has been under intense debate. I understand the passion and rationale on both sides. Yes, the plants themselves don't emit carbon, but the cradle-to-grave cycle, from uranium mining to plant construction to decommissioning and long-term radioactive waste storage isn't a carbon-free process. Then there are the safety issues. Nothing was supposed to go wrong at plants such as Three Mile Island, Chernobyl, and Fukushima, and yet it did. We live in a world where unanticipated natural disasters and power-hungry tyrants can do the unimaginable, and to me this means that skepticism about the role of nuclear power in our future is healthy.

Lee went on to fight for the Great Lakes for the rest of her career. Helping to put the brakes on the domestic nuclear industry might be enough for some people, but for Lee it was just another step on a tireless journey to protect places she loved and the larger living systems that keep them healthy. Her relationship with the Great Lakes and special places like the Indiana Dunes ran deep, and her zeal for pushing society forward was unflagging. She took on new threats to the lakes, large and small, for the rest of her life. She also brought along a new generation of emerging leaders, and her legacy lives on.

When I met her, Lee was heading up the Great Lakes Basin Commission (GLBC), based in Ann Arbor, Michigan. This was one of six federal watershed commissions established to coordinate state and federal water resources management in the nation's major water systems. Most were dealing with dams and reservoirs and focused on maintaining and delivering water supplies in their respective regions.

Lee used this opportunity to introduce ecosystem thinking and environmental quality issues into the commissions' agendas through her work in the Great Lakes. Under her leadership, the GLBC established a citizen working group to examine the role of land use—from agricultural runoff to suburban development—in contributing to Great Lakes pollution. The commission also started digging into toxic threats and potential solutions, from contaminated harbor sediments to hazardous waste incinerators. Just as these innovative approaches were beginning to bear fruit, along came the 1980 election.

The Reagan administration rapidly shut down and dismantled the basin commissions across the nation, and cast their staff and functions to the wind as part of the "government is bad and bloated" philosophy. Lee returned to the Chicago area, working at Northwestern University for a while as a research associate on Great Lakes issues and then getting deep into Chicago politics. She ran for a seat on the Chicago Metropolitan Sanitary District—a nice tidy-sounding name for the people who make big decisions about drains and sewers. It may not sound glamorous, but if you want power in water issues around Chicago, it's the place to be. She lost. She then worked for Chicago's mayor Harold Washington as head of the city's environmental agency.

All along, she remained active in multiple organizations, while writing, teaching, and mentoring folks like me. Those of us in her Great Lakes circle learned from her in subtle ways, but also through the occasional "what are you thinking?!" direct query. She raised questions, challenged strategies, offered suggestions, and engaged our brains in a way that stretched our strategic thinking. She was there in so many battles that it is impossible to name them all here.

Among the highlights beyond the work on phosphorus and thermal pollution were her decades of devotion to the Indiana Dunes—for land protection, operating dollars for the Park Service, building bridges between local communities and the steel industries, and her passion for giving young people access to the Great Lakes and opportunities to learn about them. She was with us on amendments to the Clean Air Act, the CWA, as a stalwart defender of the Great Lakes Water Quality Agreement, and much more. She received many distinguished awards and deserved them all.

At the end of the Washington administration, she shifted into private consulting and moved to the leaseback house in the Indiana Dunes that looked over a pond in the lee behind the first line of dunes on the Lake Michigan shore—a good place for a Great Lakes activist. In this modest little house with its fine many-windowed porch, Lee hosted countless gatherings with a common purpose of protecting the environment and the Great Lakes in particular. I remember several Sierra Club Great Lakes Committee meetings in which she opened her house and kitchen to a dozen or more activists. With documents and flip charts spread all over the place, we would wrestle with complicated issues late into the evening while the lake coursed into the shore just beyond the dune. Food, hearty and plentiful, would appear from Lee's kitchen to fuel the participants. We covered topics from PCBs to park budgets, and there was never a shortage of urgent priorities needing attention.

We remained friends after my career shifted away from a Great Lakes focus in the 1990s; I wasn't eager to let go of her wit, political instincts, and remarkable perspicacity. I was privileged to spend a handful of visits on welcoming summer days at the dune house. We'd walk along the beaches, taking in the view and the breeze, and as always, picking up litter—a requisite on walks with Lee.

To spend time with her in the dunes was to see a woman in her natural element. She knew the dunes in the way I knew Pine Lake. She was aware of the wind direction and what the waves would bring to the shore because of it. She knew which plants were native, which weren't, and when to sound the alarm to organize friends to yank and dig out the latter. She knew the birds and the various critters who loved to enjoy her vegetable garden despite efforts to dissuade them. She knew the unlimited spectrum of blues of the lake in its many moods, and the rise and fall of ponds and marshes with wet and dry years. She knew the place that is the Indiana Dunes.

In addition to just soaking up the place as embellished by Lee's enthusiasm, these visits included long talks on her sunny porch about politics and possibilities. She'd walk in with some tall glasses of iced tea, packed with lemons and fresh mint from the garden, and our conversations would cover topics from presidents to purple loosestrife. We'd take breaks for delicious meals that she seemed to prepare with hardly a thought or effort.

As much as Lee and her far-flung circle enjoyed that place, maintaining an aging beach house that you don't own gets to be an expensive chore. In a sense, it is also a futile pursuit when you know the NPS will eventually dismantle it and the dunes will take back the land. And, like maintaining any house exposed to the elements near the Great Lakes, it is also a lot of work. When the lease was set to expire, Lee moved into nearby Miller on the east side of Gary, to a house within walking distance of the lake and overlooking a small interdunal pond tucked into a wooded area by the elementary school. Visits there were still filled with a pleasant view—now of the little pond—and, of course, good food.

When she turned eighty in early 2008, a large group of friends, colleagues, and family gathered to celebrate Lee and her many accomplishments. She agreed to the event on the condition it would be used to raise funds for the Alliance for the Great Lakes—the organization grown from the Lake Michigan Federation she had helped found back in 1970. Those gathered feted her with witty stories and a touch of roasting. It was a grand event. As it wrapped up, she claimed the podium and warned us that her family is known for its longevity, and that she had plenty left on her agenda. She sure did, as she championed investment in the next generation of environmental leaders through the Dunes Learning Center (a youth learning program based in the Indiana Dunes), produced a television documentary, *Shifting Sands*, on the story of saving the Indiana Dunes, and continued to show up, until in her very final years, she couldn't. Lee slipped away on October 5, 2019, and I was honored that her family let me speak at her memorial, where the room, appropriately, offered spectacular views of Lake Michigan. The lady of the lakes, or "Aunt Lee," as she was often known, lived life well, and fully, and knew what it meant to be an active part of the democratic process.

Losing a mentor is hard, and I miss her. Her wisdom, tenacity, and spunk continue to inspire me, and the legacy of her leadership continues to ripple throughout the work to protect the Great Lakes.

CHAPTER NINE

The Second Wave

As a result of the first wave's activism, Lake Erie and other waters were recovering. Improved sewers, regulations on industrial discharges, and phosphate detergent bans had made a significant difference in water quality. In their book *Evolution of the Great Lakes Water Quality Agreement* (2005), Lee Botts and Paul Muldoon refer to this as the era of grappling with "visual pollution"—the damage people could easily see.[1]

Still, something was very wrong with the Great Lakes. Following the 1972 Great Lakes Water Quality Agreement (GLWQA), the IJC established a binational independent Research Advisory Board (later renamed the Science Advisory Board). Its charge was to monitor progress on water quality objectives and keep an eye out for new threats. It turned out they were finding plenty. Within a few short years, scientists confirmed the widespread presence of a troubling mix of toxic chemicals in the water and the fish. In 1978, the United States and Canada renegotiated the GLWQA, adding new language stating that the goal of the agreement was "to protect the chemical, physical and biological integrity of the Great Lakes ecosystem." It was the most ecologically aspirational agreement in diplomatic history and would drive much of the work on Great Lakes water quality for the next decade and beyond.

When I joined the Midwest office in late 1979, I knew that my boss, Jonathan Ela, had considerable experience on Great Lakes issues as well as those of the Upper Mississippi River Basin. As I settled into the role of associate regional representative, he handed me the Great Lakes issues, and turned his focus to the Upper Mississippi Basin, including the environmental impact of the river's many locks and dams. The Great Lakes portfolio opened endless possibilities for me.

My first big assignment was to secure amendments to the federal Coastal Zone Management Program to include the Great Lakes states. Success would provide access to funding for planning and restoration, originally limited to the saltwater coasts. This was my first serious foray into Great Lakes issues beyond shoreline wilderness and sand dune protection, and it was nice to start with a victory. It also required that I figure out the federal Great Lakes policy maze, in which the players included EPA, the National Oceanic and Atmospheric Administration, the U.S. Fish and Wildlife Service, the Coast Guard, the U.S. Army Corps of Engineers, the federal land agencies in the watershed, and others. Each agency has its own mission and authorities, which intertwined with those of state and local governments, tribal governments, the Canadian federal government and its agencies, and the provincial governments of Ontario and Quebec, as well as Canada's First Nations and Métis. I also learned about cross-border agreements between the U.S. and Canada such as the GLWQA, and the Great Lakes Fisheries Convention. It's quite a tangle, and the complexity of multiple layers of governance and jurisdictions continues to be one of the largest challenges to coordinate management for a system as vast and diverse as the Great Lakes.

In my early years at the Midwest office, the 1978 GLWQA was brand new. Its vision surpassed the compartmentalized regulation embedded in the growing body of domestic environmental statutes and regulatory programs in the United States and Canada. Pollution-control laws, laws to protect species and habitat, laws that regulate the shipping channels and locks, and laws that manage the fishery were (and are) all in separate silos. None had been crafted with the GLWQA's broad, integrated mandate in mind. This disparity between the domestic regulatory framework and the ecosystemic goals of the Agreement would create tensions in strategies to implement its provisions over the decades to come.

The ecological challenges that cut across the chemical, physical, and biological conditions of the lakes are not easily addressed by a single statute or policy. For example, let's look at the challenge of restoring lake trout to the Great Lakes. Initial efforts focused on reducing predation by the invasive sea lamprey, which rasps into the flesh of lake trout and sucks the life out of them. But lamprey controls, such as lampricides and weirs on spawning streams, which helped, couldn't solve the puzzle of why the surviving lake trout had trouble reproducing. Scientists suspected part of the answer lay with toxic chemicals, so this meant finding ways to reduce the chemical burdens in the lake trout food chain—from tiny organisms to prey fish. In turn, this would require chemical bans and new legislation to regulate toxic compounds coming from the air (not regulated by the CWA) and expensive harbor and sediment cleanups. Speaking of the food chain, the lake trout's prey fish were also scarce due to habitat degradation and competition from invasive species. To help the lake trout recover, we needed to address chemical pollution from several vectors, as well as habitat degradation and invasive species—old and new—and not just those that directly affected the lake trout. The puzzle required fisheries management, new pollution-control laws and regulations, as well as invasive species controls across the lakes to help restore native prey fish. It also required sophisticated research, in addition to clean, cool, suitable habitat in which the lake trout could lay those precious eggs.

This ongoing multi-decade effort has also depended on cooperation among players and decision-makers in state, provincial, Indigenous, and federal agencies across multiple departments as well as new regulations and laws. Typically, in an ecosystem, there is no single lever to pull for a simple solution. Collectively, the existing statutes did not address the system as a whole.

The 1978 GLWQA also included a commitment to address more complex and insidious threats to lakes, including toxic pollution. It set the aspiration of virtual elimination of toxic pollutants, which helped frame a concept that became known as "zero discharge"—a rallying cry for Great Lakes advocates. At the same time, the scientific and environmental communities were gaining a deeper understanding of the implications and complexity of long-lived toxic chemicals in the ecosystem. Finding responsive policy strategies to do *something* about the toxic pollution within the policy maze became a major part of my job.

Knowledge about these pollutants grew in leaps and bounds as we moved into the 1980s. This was based in part on the work of the original IJC Research Advisory Board. Rapid advances in analytical methods such as gas chromatography and mass spectrometry allowed them, and scientists around the world, to detect and identify specific chemicals, even in tiny amounts.

Among other things, the scientists were trying to understand whether there were links between the chemical structures of persistent bioaccumulative contaminants and their activity (i.e., how they behaved in the environment and in living organisms). Keywords here are *persistent* (doesn't easily break down in the environment) and *bioaccumulative* (accumulates in a living organism and/or concentrates in the food web—all the food chains within an ecosystem). The latter is more accurately described now as biomagnification.

The chemical family called polychlorinated biphenyls (PCBs) was among the most concerning on the list. PCBs are in the chlorinated hydrocarbon family, and like their cousin DDT—the pesticide responsible for widespread crashes in the 1950s and 1960s in bird populations including Great Lakes eagles—they share the property of being very long-lasting in the environment. Monsanto began to produce PCBs for commercial use in the 1930s, and PCBs were first identified as a liver toxin in the 1940s.[2]

After several decades of industrial use in high-temperature lubricants, power line insulators, and a wide range of other industrial applications, studies in the late 1960s showed that PCB contamination was widespread in the environment, including bird populations around the world. In the mid-1970s, researchers and regulators were documenting the extensive and expanding presence of PCBs in Great Lakes fish and were increasingly concerned by the potential threat to human health. As Botts and Muldoon observed, "it was realized that such stable chemicals would likely remain in the huge, virtually closed Great Lakes systems for decades if not for centuries."[3] My Canadian colleague John Jackson observed that "the very characteristics that make many of these substances valuable in a product are the characteristics that make them a disaster for the Great Lakes."[4]

When industries began to use PCBs in processes and products, their potential toxic effects weren't on the agenda. Decades later in the Great Lakes, we were learning about their potential as carcinogens (cancer-causing compounds), but the greater concern was that, like DDT, these chemicals might be reproductive and developmental toxics.

The Great Lakes advocacy community embraced the new agreement's chemical, physical, and biological framework for safeguarding the ecosystem, and toxic pollution became a priority. However, because threats to the lakes are interconnected, the agenda also included keeping an eye on pollution from routine industrial discharges, sewage, and nutrients, as well as habitat loss and wetland destruction. Some groups worked with state agencies on compliance with the young CWA; others focused on threats to fish and wildlife, from old issues such as lamprey controls, to new issues, including the proposed expansion of winter navigation through channel engineering and icebreaking. Icebreaking—especially in the St. Marys River, which connects Lake Superior to Lake Huron—was scouring the river bottom with slabs of swirling ice driven underwater from icebreaker propellors, destroying important spawning habitat. Engineering proposals throughout the system (including the St. Lawrence Seaway) also threatened shorelines, wetlands, and inshore and shallow areas. A new generation of activists rose up to take on these and other challenges. I was one of them.

By the early 1980s, there was growing consensus that by working together we could accomplish more for the Great Lakes. Once again, Tom Washington and MUCC entered the picture, backed with a substantial grant to explore founding a new, binational citizen's coalition. Wayne Schmidt, whom I'd worked with on the Michigan bottle bill initiative in 1976, led the project. In the early 1980s, he was helping lead the opposition to winter navigation. Wayne helped conceptualize the initial exploratory meeting, including its location on Mackinac Island. The island was a geographical center of the region, although a site that required planes, cars, and ferry boats to reach it. An island that doesn't allow cars does have the advantage of keeping conferees captive, but those of us who attended were highly motivated to be there because of our rising concerns about the Great Lakes.

In the weeks leading up to the May 1982 meeting, there was considerable buzz that one of the attendees was likely to be Barry Freed. The rumor circuit quickly confirmed that Barry Freed was the name under which Abbie Hoffman had been living and working as an advocate for environmental issues in the Thousand Islands region of the St. Lawrence River for some years. Abbie was a political activist and counterculture leader who helped found the "Yippies" in the 1960s. I had been in Girl Scout camp when police arrested Abbie and the other high-profile "Chicago Seven" activists during the turbulent Democratic National Convention in 1968. I always felt like I'd missed a major historical turning point, singing around the campfire while the rest of the nation watched Chicago's streets erupt into highly charged demonstrations. I had vague memories of the Chicago Seven trial, and of Abbie Hoffman, but no real basis on which to form an opinion on whether Barry/Abbie would be an asset to the coalition or not. However, his very presence at the meeting had a significant effect on the tone and outcome of the initial gathering of potential allies.

I met Abbie on the ferry going over to the island. Legend or not, he was affable, outgoing, and seemed to have a good sense of humor. He was clearly a skilled strategist, and I had the sense that he was sizing up the Sierra Club's representative to see what I and the organization might bring to the table.

We were just two among many players in this event. As memory serves, other participants included national organizations with regional offices in Great Lakes states, such as National Audubon

Society, the National Wildlife Federation (NWF), the United Auto Workers, Canada's Pollution Probe, and the Canadian Environmental Law Association (CELA). Regional groups such as Lake Michigan Federation, and Save the [St. Lawrence] River, also participated, as did numerous state and local organizations, such as the West Michigan Environmental Action Council, Save the Dunes Council (Indiana), and affiliated NWF state conservation clubs such as Michigan United Conservation Clubs and New York Conservation Clubs.

I learned a particularly important tactical lesson at this meeting. The organizers asked attendees to participate in either a working group focused on policy issues, or one focused on the structure of the coalition. I opted for the issues group because that was what I cared most about. However, I quickly learned that the power battle was being hammered out in the structure group. Often, process and structure *are* power—a lesson I rarely overlooked after that.

They were considering various organizational designs, and even at this first meeting, it was becoming clear that two models were surfacing. One was a traditional coalition structure, where various participating organizations would have a say in policy and would issue joint policy statements based on some sort of democratic process or consensus. The other model began to look more like the MUCC of Tom Washington's design: a centralized organization led by a highly visible and powerful spokesperson who would serve as the point person for moving new agendas forward.

Not surprisingly, there wasn't much disagreement about what the major Great Lakes issues were, so our "substance" group really didn't have all that much to deliberate. The process and structure issues, however, were significant, because they were going to define the power relations in the community, and frankly, we had a lot at stake.

As the discussions began to deepen about what a centralized organization would look like, and what the related implications would be for participating organizations, the momentum of the meeting began to shift from vision to details. The usual stuff of coalition building was in the conversations. Who makes decisions? Who represents whom? What happens to our existing organizational identities and issues? How do we communicate our shared agenda—through a vehicle like a common coalition newsletter or some other means? Depending on your viewpoint, we were either getting down to the tough details or getting mired in minutiae.

Abbie seized this opportunity to make what became known as the "fudge and newsletter" speech. If you haven't been to Mackinac Island, it is worth noting that the place is crawling with fudge shops for tourists. So much so, that tourists are sometimes called "fudgies," which is slightly nicer than "trolls"—a term that refers to people who live "beneath" or south of the Mackinac Bridge. Abbie tried to steer the focus back to vision and momentum for the coalition idea itself, or at least the version of it he envisioned. He didn't seem interested in sorting out the identity and power-sharing concerns of the many organizations at the meetings, which is typically part of the necessary stuff of building trust for functional coalitions. In a rousing pitch, he urged us to focus on the opportunity and challenge at hand and not get bogged down in all the questions many of us were beginning to ask about how this would all work. He admonished one suggestion under discussion: a proposal for a shared regional newsletter. He said something along these lines: "I'm not going to leave this island with just a box of fudge and a newsletter!" He already had a plan, but all the cards weren't on the table yet.

Still, most of us left that meeting energized by the potential of a larger and stronger Great Lakes movement. On the return ferry, Abbie was friendly and joking. At the little Pellston Airport, he surprised me with a kiss on the cheek and a "look forward to working with you" farewell as he boarded his flight. I don't recall leaving the island with a newsletter, but I'm sure there was a box of chocolate pecan fudge in my carry-on bag.

After that trip, I wrote in my journal: "There he was, laughing, joking, sizing up people, strategies, power blocks. His legend preceded him, and there was an air of nostalgia about his presence. The man is unquestionably brilliant, and it is almost a pleasure to watch him work with people until you realize that you are one of the people being manipulated. One thing is clear. Barry/Abbie is alive and well and working on some sort of long-range agenda. Where the environmental movement fits in is not completely clear, but we are part of his [intended] base. What is he to us?"

Abbie's involvement attracted media attention, of course. This helped raise the profile of Great Lakes issues and the need for an emerging coalition. It also raised Abbie's profile.

As discussions continued about shaping and forming a coalition, individuals and organizations began to cluster around two distinct organizational models: the strong-central-organization model and the collective-coalition model. Both can be effective structures for social change, but it's difficult to force both approaches into one structure. The strong-central-organization approach thrives on insider leverage and a certain kind of power concentrated in a master of the game and a primary player at the table. The other approach relies more on grassroots power and having diversity and breadth in multiple bases of power and influence all at the same time. In partnership, these two models of collective leverage can be highly effective, providing something of a "one-two punch" in political action. But a hybrid within a single structure is almost counterintuitive because the methods and the tactics are so different.

With experience in the successful Alaska Coalition, I was leaning toward the model based on optimizing the power of existing organizations, as were many others. Abbie and his organization, Save the River, formed the heart of the other block, along with several NWF affiliates, including MUCC, as well as the UAW. The divergent views generated growing friction as we approached the emerging coalition's next meeting, in which we were to adopt bylaws that would define our structure and operating rules. This time around, I was paying closer attention to structure.

I soon learned that having a different opinion from Abbie Hoffman in matters of strategy could be costly. In the days leading up to that meeting, Great Lakes reporters were picking up on the tension about what shape this new coalition would take. A *Chicago Sun-Times* reporter called me and wanted my response as to whether I felt the success of a new coalition depended on a strong, charismatic new leader in the Great Lakes community. I suspected he was testing the waters about Abbie in that role, but I was disinclined to have this discussion in the press. After the reporter pushed me on the topic from a different angle, I made a statement to the effect that I didn't think we needed a Great Lakes czar, never mentioning anyone by name.

Apparently, before the paper had gone to bed, the reporter had called Abbie to let him know what was in the story. The next day I received a call from Mike McCloskey, executive director of the Sierra Club. As a lowly associate field representative at the time, I had little routine contact with the executive director, so you can imagine my surprise at his call, and then the subject. Abbie was so outraged at my oblique reference and apparent questioning of his preferred strategy that he had demanded that the Sierra Club fire me. Mike, to his credit, simply wanted to hear what I had to say and found my actions reasonable and not worthy of Abbie's fury. I kept my job, and my defensive radar went on full alert. The guy was brilliant, but vindictive—and I could see that a fair debate about organizational structure at the upcoming meeting was not in his game plan.

The organizers held the big meeting in a building on the University of Windsor campus in Windsor, Ontario. It was certainly one of the most bizarre environmental meetings in my entire career. There are many versions of what took place at the Windsor meeting. Here's mine.

Those favoring the member-led coalition included me, and representatives from Pennsylvania's NWF affiliate, the National Audubon Society, Lake Michigan Federation, Ontario's Pollution Probe, and others. Abbie had brought a large team with him from Save the River to push for the strong-central-organization model. Allies still included some of the sportsmen's groups and the UAW representatives.

Everyone was jumpy. A colleague advised me to avoid conversations in the restroom because they had learned that Abbie's team had assigned someone to listen to what I was saying and report back to Abbie. This was just too strange to believe. But considering the week I'd had with Abbie's call to Mike, I kept my mouth shut.

At one level, this was organizational dysfunction run amok. At another, it demonstrated how important the Great Lakes were to all of us in the room, that we thought it was worth struggling through all this internal power stuff to build a stronger base of action for the lakes. By the end of the first day, several of us headed out of the meeting site to clear our heads, get away from the "listeners," and talk over strategy at a charming Ontario restaurant called Ducks on the Roof.

While dining on duck, we mapped out a plan to introduce amendments to the proposed bylaws that would ensure a large decision-making role for participating member organizations, which is what we wanted in the first place. We'd been accused earlier that day of making a power grab for national organizations by resisting the model that Abbie was proposing, which was clearly based on a high-profile leader with largely independent decision-making power. I found this laughable. If only our opponents had realized just how hard it was for those of us in the field to get our national organizations to pay attention to Great Lakes issues in the first place, they would have known how far off the mark their accusation was.

On the second day, everything hit the fan. We introduced our primary amendment to shift the decision-making power (in terms of policy direction) to members of the coalition, and during the introduction, a long group of allies lined up behind the microphones to speak in favor of it (a tactic I had learned from that MUCC convention years before). It was clear we had pulled a lot of votes in this direction. Abbie took this debate as the opportunity to force the unspoken question. I honestly don't remember much about his speech, other than he was livid. He announced that he and the delegates from Save the River were walking out, and the UAW was walking out with him, which, of course, would have been a crushing blow to the coalition.

This was a tactical error. He'd neglected to clear this with the UAW representatives first. Russ Gossman, UAW's conservation director and a tough-minded but genial guy who I'd worked with in the battles to pass the Michigan Wetlands Bill, rose to his feet. "No one," he said, "tells the UAW when to walk except the UAW." We quickly realized they had no plans to walk out. Abbie's only option at that point was to carry out his threat to leave, and after a stunned pause to recover following his dramatic exit, the meeting went forward, albeit awkwardly, with the UAW and most of the other participants still in the room.

In something of a tragic victory, the final bylaws we cobbled together in the conciliatory aftermath of that scene were indeed a hybrid. At the end of that exhausting meeting, New York Conservation Club's Bill Boyce rose to spread a little oil on the troubled waters and offered "Great Lakes United" (GLU) as a name for the coalition. It passed with enthusiasm.

The new bylaws allowed member organizations to provide direction on GLU's policy agenda via resolutions, adopted by coalition members at annual meetings (with provisions for urgent actions between meetings). The bylaws also left open the option of hiring a director and other staff. This policy process was a bit cumbersome, but definitely participatory.

Soon, the member groups in the young GLU agreed that we needed an executive director to coordinate the organization's efforts. Dave Miller, its first director, was a smart choice. With his leadership, in the 1980s, this coalition rose to help drive some of the most significant Great Lakes protection legislation in the history of both nations—toxic pollution controls for air and water, cleanup strategies for toxic hot spots, and funding for myriad research and restoration endeavors. In retrospect, it was worth all those hours in ghastly meetings. Save the River ended up playing a role in the coalition on St. Lawrence River issues after all, although Abbie's presence was never again front and center in GLU gatherings.

Abbie went on to fight other battles, taking on other aspects of "the system" for the rest of his life. We later learned that he suffered from bipolar disorder. He died from an overdose of phenobarbital pills in 1989 in what the media reported as a suicide. Many who were close to him could not reconcile this conclusion with the man they knew. Leadership takes on many forms, and he was a remarkable leader in his own fashion. I was sorry to learn of his struggle with bipolar disorder, and his death.

The Great Lakes movement was, and is, larger than any individual, and that is the strength in real movements. Playing well with others is never easy when passions, power, and money are at play. Various Great Lakes coalitions and alliances have changed in shape and configuration, and leaders have come and gone over the years. Political bedfellows come together out of common need or opportunity, but they don't necessarily sing in close or comfortable harmony. Many people and organizations want and need the Great Lakes to be cleaner, healthier, and more vital. That aspiration brings people and organizations together, even across philosophical and strategic divides.

By the mid-1980s, GLU took its place among a team of dedicated Great Lakes leaders. This included staff in the field offices of U.S. national organizations such as the National Wildlife Federation, the National Audubon Society, and the Sierra Club, as well as Canada's Pollution Probe and CELA. Leading regional players strengthened this alliance, including Citizens for a Better Environment, Lake Michigan Federation, and Quebec's Société pour Vaincre la Pollution, along with many state environmental councils—and yes, MUCC, Save the River, and the UAW in addition to its Canadian counterpart. The wave was building, and our power was growing in Washington, Ottawa, and the capitals of the Great Lakes states and provinces. Regardless of our internal struggles, we were indeed united by a passion for the lakes, and we had much to accomplish.

CHAPTER TEN

Chemistry, Biology, Urgency

With the coalition-building hurdles behind us, the Great Lakes movement geared up to pursue big opportunities. Foundations that had supported organizing GLU were also funding complementary efforts, which gave the Sierra Club the boost to develop our Great Lakes program. Networks we developed through the efforts to organize GLU were coalescing in informal teams on key issues. I joined a group working on toxic pollution, which included the Sierra Club, Citizens for a Better Environment, Lake Michigan Federation, National Wildlife Federation, Pollution Probe, and the young Great Lakes United.

We also collaborated with the region's scientific community, including the IJC's Science Advisory Board (SAB) and academic and government scientists, many of whom were members of the International Association for Great Lakes Research. Through reports from the SAB, the Great Lakes Water Quality Board, and papers emerging from the field, public agencies and Great Lakes advocates were continuing to learn about the extent and impacts of toxic contamination in the Great Lakes food web. Driven by the

Great Lakes Critical Pollutants (1987)

CHEMICAL GROUP	ORIGIN/USE	PATHWAYS
CHLORINATED HYDROCARBONS: ORGANOCHLORINE PESTICIDES		
Aldrin/Dieldrin	Pesticide	Application sites, runoff, manufacture, atmosphere
DDT (and metabolites DDD, DDE)	Pesticide	Application sites, runoff, manufacture, atmosphere
Hexachlorobenzene	Fungicide and by-product of chloralkali and wood preservative manufacture	Application sites, manufacture, incineration, atmosphere
Mirex	Pesticide	Application sites, runoff, manufacture, atmosphere
Toxaphene	Pesticide	Application sites, runoff, manufacture, atmosphere

aspirations of the 1978 GLWQA, the SAB developed a list of the most worrisome toxic pollutants in the Great Lakes system, which many of us called the "nasty eleven." It provided a focal point for our efforts in the late 1980s and beyond.

I've organized the "Great Lakes Critical Pollutants (1987)" table by chemical "family." One quick note: living things, from bacteria to humans, endeavor to break down or isolate and store toxic chemicals we ingest, inhale, or absorb. This happens through our own immune systems and biochemical metabolism (the process by which we get energy from our food). Chemicals called "metabolites" are results of the biochemical breakdown process, some of which are equally

CHEMICAL GROUP	ORIGIN/USE	PATHWAYS
CHLORINATED HYDROCARBONS: INDUSTRIAL CHEMICALS AND BYPRODUCTS		
PCBs	Over 300 industrial uses (high-temperature lubricant)	Wastewater, products and disposal, atmosphere
PCDD (dioxins)	Industrial processes or organic combustion by-products	Wastewater discharge, incineration, atmosphere
PCDF (furans)	Industrial processes or organic combustion by-products	Wastewater discharge, incineration, atmosphere
METALS AND METAL COMPOUNDS		
Alkyl-lead	Lead compounds used primarily as gasoline additives	Auto exhaust, gas stations
Mercury and mercury compounds (e.g., methyl mercury)	Natural element and metabolites, industrial uses, and waste streams	Wastewater, products, mining, coal/ oil combustion, incineration, atmosphere
POLYCYCLIC AROMATIC HYDROCARBONS		
Benzo(a)pyrene	By-products of incomplete combustion of organic matter	Coal/oil combustion, incineration, atmosphere

toxic (or more so) than the original compound. Thus within the food web (the interconnected food chains within and around the Great Lakes), the organisms that contain breakdown compounds are also a pathway for some chemicals.

Let's take a quick walk through these chemicals and their properties. Remember that early investigation into chemical structure? The scientists learned that when you marry chlorine to benzene rings or carbon chains with a covalent bond, we should pay close attention to how they behave in the environment. The compounds containing chlorine and carbon are called chlorinated hydrocarbons. They are robust and persistent, meaning they don't easily break down in the environment or go to the magical place we call "away."

Chlorinated hydrocarbons include PCBs, DDT and its metabolite DDE, as well as several other now-banned or highly restricted pesticides, as well as dioxin and furan compounds.

Pesticides were supposed to kill pest insects and make your life better. Eradicating pesky insects was also the job of dieldrin (for fruit pests). The same goes for toxaphene (for boll weevils in southern cotton crops) and mirex (for fire ants in Texas and the South, and as a fire retardant). Seed producers used hexachlorobenzene as a fungicide to treat seeds for crops such as wheat. Organochlorines were never intended to accumulate in the food web or recalibrate the internal biochemistry of birds to thin their eggshells and prevent successful incubation, but they did. They were never intended to show up in our bodies either, but they did. By attempting to find a quick chemical fix to control undesirable insects, we ended up messing with Mother Nature and ourselves as well. As Jane Goodall observed: "How could we have ever believed that it was a good idea to grow our food with poisons?"[1]

You will not find the 209 synthetic (human-made) chemicals that comprise the PCB variants in nature unless we put them there. For the chemistry curious, "polychlorinated" means the chemical structure of these compounds has two or more chlorine atoms. "Biphenyl" means these compounds have two benzene rings connected by a single covalent bond. If you replace hydrogen atoms in those benzene rings with chlorine atoms, you get PCBs. This structure matters because that's what makes them very robust in the environment and, unfortunately, also a good fit with many receptors in living cells in vertebrate bodies (that includes us). I won't go into the deeper weeds of biochemistry here, other than to say it is the presence of chlorine (or other halogens) and the chemical structure that enables these compounds to mess with hormones that affect reproduction, as well as physical, emotional, and intellectual development.

The mounting evidence of damage from these chemicals led both the United States and Canada to ban or restrict the use of PCBs and most of these pesticides. In some cases, such as DDT, manufacture continued for export. Today, North Korea, India, and China still manufacture and use DDT. Traces ride across the world on the winds of the atmosphere, and we now find "new" DDT in the Great Lakes. In living organisms, DDT breaks down into DDE—and that's just one of its breakdown compounds. Unfortunately, DDE and many other chemical breakdown compounds from banned chemicals are also very nasty food web contaminants. Unlike regulating a manufacturing process, we don't have the means to ban something produced by biological metabolism in living organisms.

We intentionally made PCBs and these pesticides (including herbicides that contained furans). We never sought to produce dioxins and furans that are by-product contaminants from manufacturing processes that use elemental chlorine, such as pesticide manufacture or some paper bleaching processes at paper mills. Combustion of chlorinated compounds and plastics (e.g., polyvinyl chloride, a.k.a. PVC) can release dioxin, and medical or municipal waste incinerators can be big sources. Dioxins have been described as "one of the most notorious anthropogenic environmental toxicants," and are one of the world's "dirty dozen" persistent organic pollutants.[2] The worst dioxin compound is something named 2,3,7,8-TCDD, which was a contaminant in the banned chemical defoliant Agent Orange. Assassins used dioxin compounds to try to kill Ukrainian president Viktor Yushchenko during his 2004 presidential bid. He survived, albeit scarred and disfigured from "chloracne," a manifestation of dioxin poisoning. Furans are also combustion and process by-products, and very similar to dioxins but with a slightly different chemical structure. You don't want any of these compounds in your food web.

As noted, the structure of chlorinated hydrocarbons means they don't break down *easily* in the natural environment. Some do break down during metabolism, or exposure to heat, or ultraviolet light, but *slowly*. Chemical structure is also why oil and water don't mix. Chlorinated hydrocarbons fall on the oil-loving side of that divide. The scientific word for this is "lipophilic," a Latin construction meaning "fat loving." This divide is important because it tells you where they are likely to end up in a fish or human and indicates how they behave in a food web.

Lipophilic toxic compounds accumulate in the fats and oils in living tissue. When one critter consumes another, it absorbs that creature's entire storehouse of these chemicals. As contaminants move through a food chain, from the tiny organisms up to the predators, the concentrations magnify at each stage of the food chain, which gives us the term "biomagnification." Predator fish (who eat smaller fish, who in turn ate their share of other organisms) that are also rich in natural oils, such as salmon, lake trout, or tuna, are the ideal places for these chemicals to accumulate at the highest levels of magnification in their food chain. Trimming the fat and skin from any fish with a suspected body burden of chlorinated hydrocarbons helps reduce your exposure (although the fish never has this option when it is dining). I'll leave you to figure out what to do with the skin and the trimmings, but don't feed them to your cat.

Chlorinated hydrocarbons are just one group of halogenated hydrocarbons. You may not have looked at a periodic table since high school chemistry, but fluorine, chlorine, and bromine are neighboring elements. When combined with a robust hydrocarbon structure, fluorinated, and brominated compounds are also persistent in the environment. Examples of halogenated hydrocarbon contaminants include per- and polyfluoroalkyl substances (PFAS) and perfluorooctanoic acid (PFOA), now nicknamed "forever chemicals" owing to their persistence. Bromine compounds gave us polybrominated biphenyl (PBBs), an infamous chemical that contaminated Michigan animal feed in 1973.[3] The images of the deformed calves and trembling cattle remain in my memory, and I assume that traces of the compound are still in my body, and in thousands of others who were consuming milk, eggs, and meat from Michigan in 1973.

Concern about the adverse health effects of PCBs included its potential to cause cancer; immune suppression; reproductive, nervous, and endocrine system damage; and more. This led the United States to ban further PCB manufacture in 1979 as part of the Toxic Substance Control Act, following Canada's lead in 1977. Unfortunately, when the toxic cat is out of the bag, bans don't have the capacity to mop up and contain what is already in the environment. The research community knew one repository of PCBs in the Great Lakes was the sediments that had collected in Great Lakes rivers and harbors downstream from industries that had used or manufactured PCBs. In 1980, EPA's measurements from Lake Michigan's Waukegan Harbor sediments found some samples were almost 50 percent PCBs.[4] Scientists also knew the sediments weren't the only pathway to Great Lakes waters or wildlife. In 1976, Dr. Wayland Swain and his colleagues found PCB concentrations in the flesh of wild lake trout in Siskiwit Lake on remote Isle Royale in northwestern Lake Superior.

These chemicals had found their way into what Dr. Swain (and others) described as the largest lake on the largest island in the largest lake in the world. Not only had they reached the isolated corners of Lake Superior; PCBs had also efficiently concentrated to measurable levels in lake trout throughout the lakes. In 1980, scientists found the boll weevil pesticide Toxaphene in the Siskiwit Lake fish as well, far from the cotton-growing regions of the southern United States. They also found traces of other chlorinated hydrocarbons, including the pesticide Lindane, as well as furans and dioxins. Studies in Lake Ontario found Mirex, which had been banned since 1976. This research confirmed that toxic pollutants were traveling through the atmosphere and deposited through rain, snow, or dust.

Efforts to revive the lake trout population were failing in most Great lakes, even with effective lamprey controls. Data pointed to the toxic chemicals as one likely cause of reproductive failures. In addition, fish-eating birds like bald eagles were still having trouble reproducing, long after the domestic DDT bans.

Over the 1980s, scientists found more troubling evidence. Fish-eating birds such as terns and cormorants had abnormal rates of birth defects and failed hatches in the islands of northern Lake Michigan. Scientists, including Jim Ludwig and Tim Kubiak, documented these malformations and egg failures in Great Lakes cormorants. They looked for linkages between the effects and the presence of PCBs and other chlorinated organic compounds. Like Great Lakes eagles, fish was the birds' primary food source.

Research also found that the PCBs in sediments weren't staying put. Biological activity among benthic (bottom dwelling) organisms, along with the presence of volatile compounds within the sediments, meant bioaccumulating chemicals were percolating up the water column into the atmosphere where wind currents, rain, and snow would redeposit them in other areas. One study determined that the largest source of new PCBs in Lake Superior was coming from the old PCBs in Lake Michigan.

Beyond the halogenated hydrocarbons, heavy metals were also of concern—most notably alkyl-lead and mercury. EPA had already initiated efforts to reduce lead contamination by restricting alkyl-lead compounds in gasoline in 1973 (finally banned in 1996). Unlike chlorinated hydrocarbons, heavy metals have no affinity for fatty tissue, but they do bioaccumulate and some biomagnify. Our bodies absorb them and try to eliminate them in many ways. For example, lead moves into bones, kidneys, and other organs through the blood, and even into the brain. Our bodies try to store it out of the way in bones and teeth, but it can be released during stresses on the body. Its capacity to damage developing brains is the primary threat to children living in houses that still have lead-based paints or drinking water contaminated by lead pipes—still a challenge for many communities in the Great Lakes region.

Mercury was of greater concern, as there were few controls and it accumulates in the aquatic food web. Once again, atmospheric sources appeared to be playing the largest role. Mercury—or more specifically methylmercury, a very toxic compound formed when the element itself interacts with natural bacteria in the environment—accumulates throughout the flesh of a fish. Trimming the fat doesn't help if you are trying to reduce your mercury intake. Given its toxicity and ability to mess with our neurological system, behavior, gene expression, reproduction, and fetal development (and just flat out kill you at high doses), reduction seems wise, which means eating less fish, or lower on the food web, or less often, or all three.

Polycyclic aromatic hydrocarbons (PAHs) are another family of chemicals that got the SAB's attention. Near industrial areas, and especially near steel plants and coke ovens, anglers were finding bizarre tumors and lesions on fish, and PAHs were the likely culprit. PAH compounds are formed by the incomplete combustion of organic material, whether that organic matter is coal, oil, wood, tobacco, or other combustible materials containing carbon. The SAB specifically flagged benzo(a)pyrene, a known carcinogen and one of the most potent PAHs found in (but not limited to) cigarette smoke.

You aren't getting much of a dose from an occasional fire in the fireplace on a winter's day. However, if you work at a steel mill with a big coke oven, live downwind, and smoke cigarettes, please change at least one of those things. PAHs are widespread lake contaminants, typically concentrated in nearshore industrial sediments. Although they accumulate in tissue, they aren't as efficient at it as the halogenated hydrocarbons. Their primary toxicity endpoint is cancer. Physical exposure to the PAHs in bottom sediments was the trigger for the tumors and skin lesions in fish. Our metabolism can also produce more toxic forms of these compounds (like what happens when our gut biome converts the nitrates we eat in bacon into more troubling nitrites). Some PAH metabolites are potent mutagens.

Armed with this growing body of knowledge, Great Lakes advocates set out to reduce loadings and exposure through responsive policy. The first generation of pollution-control laws weren't designed for toxic chemicals building up in the food web. The 1972 Clean Water Act targeted sewage, industrial waste, and nutrients. Food web contamination and atmospheric transport were barely on the radar then. Further, the law wasn't designed to deal with toxic chemicals in products that may be released during use, recycling, incineration, or disposal. And, for persistent bioaccumulative chemicals, its dilution-based approaches don't really work—especially in a system where the water and sediments remain for decades and even centuries.

We faced a big public learning curve to gain support for new policy. Congress passed the National Cancer Act of 1971 as part of the War on Cancer, and it fueled significant new national investments in cancer research. The question "does it cause cancer?" helped shape the investigations into toxic chemicals as well as the regulatory process. However, cancer is not the sole effect of most toxic chemicals, and when you are primarily looking for carcinogenic triggers, you can miss other signs of trouble. To build awareness and dialogue about what the scientists were finding, we issued reports, filled our newsletters with articles, hosted talks and workshops, and lobbied. A common question was, "is the water safe to drink?" In that era, if your water supply came from the Great Lakes, the answer was generally "yes," given that concentrations in the water itself were exponentially less than those accumulated at the top of aquatic food chains. It gave us a chance to explain the impact of bioaccumulation and biomagnification by stating that you'd have to drink two hundred thousand glasses of Great Lakes water to get the same dose as consuming a meal from one large lake trout.

In workshops, I watched Wayland Swain explain the concept of the "half-life" of chemicals stored in our bodies with the clever use of volunteers and a clothesline. He'd stretch out the clothesline with a volunteer on each end, and explain that it represented a body burden of heavily chlorinated PCBs. Then he'd fast-forward the timeline twenty years and cut the clothesline in half, taking half away, leaving the volunteers to hold the remaining half. As the timeline moved forward another twenty years, another half would be gone, but they kept the remaining quarter, and so on. While the body burden in the demonstration slowly got smaller, it took a very long time. Wayland was never shy about telling people how serious the contaminant issues were, and he earned the nickname "Dr. Doom." His Q&A sessions were always illuminating. When someone tells a woman her current body burden may be traceable (if each generation nursed their female offspring) through her daughter, granddaughter, and potentially beyond, it makes you think. We wanted our policymakers to think too.

CHAPTER ELEVEN

Cosmos Goes to Washington

To build momentum for action on toxic chemicals and other Great Lakes issues among decision-makers, the Sierra Club and GLU teamed up to host a series of annual Great Lakes Washington Weeks, beginning in 1986. These events brought around twenty volunteers and staff from U.S. and Canadian environmental and allied organizations to Washington, D.C., for an intensive week of learning and lobbying. We knocked on just about every door of Congress or agency that had jurisdiction over some aspect of Great Lakes decision-making.

One of my favorite meetings was at EPA headquarters, in part because they just didn't know what to do with us. The Washington-based staff of this agency rarely interacted directly with grassroots leaders, and we brought them twenty passionate and articulate Great Lakes folks who wanted to talk over serious issues, not tour the headquarters. We didn't look or act like the usual corporate lobbyists who often frequented their halls. My efforts to schedule the meeting illustrated the gap between the structure of environmental regulation in the United States and the ecological realities we were facing in the lakes.

The circular conversation with the person who took my first call to set up the appointment went something like this.

> Me: "We want to talk with key agency staff about how toxic chemicals in the air are polluting the water and contaminating the fish in the Great Lakes food web."
>
> EPA: "So do you want to meet with staff in the Air and Radiation Division, or toxic substances, or the people who work on the Clean Water Act regulations?"
>
> Me: "We want to meet with them all, at the same time. It's all connected."
>
> EPA: "But they're on different floors in different departments, and we'll have to clear this through several department heads. Maybe it would be more efficient just to start with the Air and Radiation Division."
>
> Me: "Are they dealing with contaminated fish?"
>
> EPA: "Not to my knowledge. That would probably be FDA."
>
> Me: "We'd like to schedule a meeting with all the relevant EPA divisions that can help us address toxic air pollution that is polluting the water and contaminating the fish."

Eventually we secured the meeting, but it was sobering for some in our group to experience what those of us who'd been at this awhile knew all too well. The job of the Environmental Protection Agency is to enforce the nation's environmental laws as written, and when you present a problem that falls through the cracks of the body of existing law, they don't have a mandate to initiate an innovative approach unless Congress tells them to. I remember one volunteer saying, "But I thought their *job* was to protect the environment." Well, it is, sort of—as far as the law and the compartments of the agency allow.

Unfortunately, our laws force complex environmental threats into compartments in distinct divisions managed on different floors. This doesn't reflect the ways in which gravity, chemistry, physics, biology, and the landscape interact. We'd brought a new generation of issues to the old-school way of managing the environment, albeit the old-school laws like the CWA were only a dozen years old at this stage—infancy in the body of law.

However, from the rise of early concerns about pollution in the 1960s, human understanding of the living environment and its complexity had grown in leaps and bounds over those two decades—especially in the Great Lakes Basin, where the binational research efforts were driving new discoveries. As a result, we faced a growing disparity between the framework of law and regulation and ecological reality.

We began to push the envelope and ask for new approaches to things that couldn't easily be managed by "end-of-the-pipe" solutions and a CWA discharge permit. These permits, by design, are "technology-based"—in other words, what does the best available technology enable you to capture or prevent from discharge? If the technology alone can't meet ambient water quality standards for the receiving water body, the next regulatory tool in the hierarchy is reducing risk (which is different from prevention of harm). Risk-based standards allow industrial plants or sewage treatment plans to discharge certain amounts of specific pollutants into lakes and streams. The risk calculation (for human health and the environment) factors in the dilution capacity of the receiving water body. EPA regulates toxic pollutants on a substance-by-substance approach that doesn't factor in cumulative risk over time, or from multiple sources in the same waters, or combined risk from exposure to multiple chemicals. It might be functional from an administrative standpoint, but it is not a fully effective approach to keep our waters clean and healthy for humans and other life.

These National Pollution Discharge Elimination System permits (NPDES), known in the trade as "NIP-e-dees" permits, apply only to a specific point of discharge (such as an industrial wastewater pipe) and are thus known as "point sources." The CWA doesn't cover diffuse sources, known by the clumsy term "nonpoint." They include sources such as fertilizer and pesticide runoff from farm fields; groundwater contamination from old industrial sites, landfills, or septic tanks; and airborne deposition from the atmosphere.

The word "permit" itself points to part of the problem. In one light, it is a noun for the license to discharge wastewater at certain amounts into a river or lake, as in "we received our discharge permit." But the word is also a verb. It permits, allows, or even sanctions that industries and municipalities can discharge some regulated pollutants into waterbodies. For chemicals that are robust in the environment, or for troubled water bodies burdened with nonpoint pollution, just treating water up to the standards on the books can permanently normalize degraded conditions. Moreover, for chemicals for which standards don't exist, until trouble manifests, they go unregulated, as was the case with the PFAS that we are now wrestling with in the 2020s.

So, we were trying to force a dialogue among policymakers and regulators about the need to connect the dots between *sources* of pollution, *how* contaminants are dispersed in the open environment, and their direct and indirect *effects* on living organisms—many of which were more subtle than acute illness or death, but still very concerning. Because Congress didn't design the laws to deal with these effects and their causes, it was difficult to make our case among those charged with administering the existing framework.

In an extraordinary Washington Week in 1988, we threw a Capitol Hill reception to help illustrate our points. Hill receptions are so commonplace that starving interns and junior staff can sustain themselves fairly well by "grazing" the various receptions that their bosses get invited to all the time. There's usually food and drink, a lot of handshaking and exchanging of business cards, elbowing to connect with the important attendees, and an obligatory speech with polite applause muffled by the attempt to balance wine glasses and a tiny plate full of hors d'oeuvres at the same time. I grew to dread receptions.

We decided to throw one with a twist. We enlisted the help of "Scootch" Pankonin, a successful contract lobbyist known for the great parties she hosted at her Capitol Hill townhouse. Scootch was a phenomenal cook and often prepared many of her own wonderful spreads. She had a soft spot for the environment and in a weak moment agreed to prepare the Hill's first "toxic buffet" for Great Lakes Week. A Great Lakes congressman secured a meeting room for us in the House Office buildings and invited the staff and members from all the House districts and Senate seats in the eight Great Lakes states, along with agency staff, friends from the Canadian Embassy, other colleagues, and Washington-based reporters for the region's news media.

The twist was that every morsel of food, every drop of water, and every drink at the bar came from the waters of the Great Lakes ecosystem. We served the top catch of the Great Lakes: lake trout, Chinook, and Coho salmon. The Akwesasne Nation even donated a prized sturgeon for the event. Volunteers from the region flew in with gallons of their tap water, and a few were gracious enough to ship an iced frozen salmon or trout in their checked baggage as well. Imagine what today's airport security people would do with a gallon of Lake Michigan tap water and fish whose fat layers may have constituted hazardous waste. Organizing was simpler back then.

Scootch prepared court bouillon and poached these beauties in her kitchen, presenting them on dill and lemon-bedecked platters on our buffet table. The elegant presentation attracted our guests, but a card that accompanied each fish included the warning information from the applicable state's fish advisory booklets. The cards noted the likelihood of the presence of PCBs and other contaminants in the flesh of these fish. These warnings advised children and women of childbearing age not to eat fish of this size and species, and in some cases the advisories applied to adult men as well.

Our centerpiece on the buffet table included Henrietta, a taxidermied cormorant provided by Tim Kubiak, the scientist who had found her. She had a recognizably deformed bill that the scientists attributed to the chemical contamination in the lake food web. The bar featured five pitchers of water—one filled with drinking water from each of the Great Lakes—along with New York and Michigan wines, and beer from the region's breweries. The Wisconsin cheese and crackers provided most of the sustenance that our guests dared to eat at the party. The television reporter from a Duluth station felt obliged to sample the Lake Superior lake trout on camera in an act of regional pride, but the rest of the fish was untouched as people read the warnings and glanced at the cormorant.

While our guests on the Hill declined to sample the catch of the lakes, tens of thousands of anglers and their families in the region were eating similar fish all the time. Our point was that Congress was in a position to lead efforts to protect the lakes from further contamination and clean up the waters and thus the fish. Saying "no thanks" to the salmon was a first step in getting members of Congress and their staff to pay attention to this issue. Our hope was that they would eventually say yes to getting the chemicals out of the fish their constituents ate.

The following year, we returned to the Hill, this time with Cosmos, a live cormorant that Jim Ludwig brought to Washington in a large crate-sized box in his car. Like the previous cormorant that had graced our buffet table (albeit posthumously), she had an equally impressive twisted-bill deformity. Jim was hand-feeding her at his home in Michigan to keep her alive. She also had a lively personality that made her hard to ignore. She was part of the living evidence in a hearing on toxic air pollution as the House considered amendments to the Clean Air Act. We barely got this large bird through security, but we did. I'll never forget her sunning her wings in the heat of television camera lights as members of Congress and committee staff stared into the face of what the chemicals of our times were doing to Great Lakes wildlife. A large, squawking black cormorant in a formal hearing room is a hard thing to ignore. She may be the only cormorant in history to have "testified" in front of a congressional committee, or to have graced the pages of *Life* magazine as a result. She did a great service for her fellow creatures in the Great Lakes.

During this rising tide of attention to Great Lakes issues, the Water Quality Agreement came under review again. The parties were considering substantive changes that would require formal negotiations between the two nations. Given the level of citizen activity on both sides of the border, the agreement was under some of the most intense public scrutiny in its history. In 1987, EPA invited me to serve as one of three U.S. "citizen observers" in the negotiating process, along with Great Lakes United's Tim Eder and National Wildlife Federation's Mark Van Putten. In the observer role, we were allowed to be present during the formal negotiating sessions and on some of the planning calls, and to serve as advisors to the EPA and State Department team at their request, but we had no direct power or authority.

The Canadians set up a similar team, which included Kate Davies, from the city of Toronto's public health department, and John Jackson, who was on the board of GLU. It was a privilege and responsibility to serve as an observer. I have vague recollections of swearing an oath to be loyal to my nation as we joined the process. None of the observers considered this a passive role, and each of us did intense homework to bring whatever substantive knowledge we could to the process related to our primary concerns.

As such, in the U.S. team, within our stated roles, Mark took a lead on matters that dealt with overall water quality, including goals and standards; Tim covered the bases on contaminated sediments; and I served as point person on air toxics. The U.S. EPA took the lead on developing the substantive points in the U.S. position, but the State Department was the boss on diplomacy. The parallel regulatory and diplomatic agencies in Canada—Environment Canada and the Department of External Affairs—played similar roles on the Canadian team.

It was fascinating to witness the effort to craft something whole and meaningful for the Great Lakes out of environmental reality, existing regulatory structures, and the Reagan administration's limited enthusiasm for environmental protection. The teams wrestled with substance, process, and the tools of diplomacy. There was a dance of ideas as the observers and members of the negotiating team made efforts within our constraints to advance key provisions to protect the lakes and respond to pushback from the Reagan White House, EPA, or the Canadian team. We attempted to artfully thread the procedural needles in creative ways to weave the best substantive improvements we could. Any recommendations that might ruffle feathers or push the envelope meant we also needed defensible rationale and/or political coverage—especially for those who had more than a substantive stake in the negotiating process, whether those stakes were serving political loyalties or safeguarding careers.

During the deliberations, we almost lost the entire section on toxic air pollution due to White House concerns about blurring the issue with acid rain—a sensitive topic between Canada and the Reagan administration at the time. Acid rain was the simple term that described the acidity in rain and snow caused by sulfur dioxide (SO_2) and nitrous oxides (NO_x) in the atmosphere. The compounds came primarily from coal-burning power plants. Acid rain was changing inland lake water chemistry to the point where fish couldn't survive. Much of Canada's acid rain problem was caused by emissions from the United States—primarily from the Ohio Valley and Eastern power plants. Canada wanted U.S. action; the Reagan administration wanted to protect the coal industry.

The atmosphere disperses and deposits toxic and acidic compounds in a similar fashion, although from diverse sources. Our toxic substances agenda in the agreement didn't include the acidic compounds. Still, this annex was giving someone in the Reagan administration pause. The working U.S. draft of the air toxics annex was titled "Airborne Toxics." The Canadian version was "Atmospheric Pollution." This then morphed into "Airborne Toxic Pollutants" in the first joint draft.

This language was apparently too close to the terminology around acid rain for comfort in the Reagan administration. Nine days later, during the final hours of negotiations, the entire annex was at risk. Val Adamkus, head of EPA's Region V, and a leader for the U.S. negotiating team, was on the phone with someone we presumed was in the White House. Val paused the vigorous conversation long enough to ask those of us in the room for an alternative title to save the annex. I grabbed a pad of paper and rushed to write several options to describe toxic air pollution that might fly under the radar of acid rain anxiety. I handed him several different phrases, while he was providing assurances to whoever was on the other end of the line. He shared some of the alternative phrases, and after a few tense moments, he nodded. I exhaled. Thus, "Airborne Toxic Substances" Annex 15 became part of the 1987 protocols. Such is the art, craft, and drama of diplomacy.

The substantive improvements from this negotiation were due in part to the dedicated work and cooperation among unlikely allies on both teams. Several new annexes were among the greatest gains. Annex 14 established a framework for remedial action plans to clean up the forty-three contaminated sites designated as "Areas of Concern" (AOC) in the Great Lakes Basin. Annex 15 took on airborne toxic substances as noted. In addition to the new annexes, the new language also included Appendix 1: "Hazardous Polluting Substances," which lists 272 substances of known concern in the Great Lakes, and a second appendix with 108 more *potential* hazardous polluting substances. It felt like we had achieved something significant. This fresh version of the agreement was technically known as "The Great Lakes Water Quality Agreement of 1978 (Revised) as amended by Protocol signed November 18, 1987."

The observers for both nations shared a mutual respect for the process and the outcome. However, while we focused on substance, the agency people on both sides of the border were also tinkering with structure and process. Once again, that stuff can come back to bite you.

We left the final session on the agreement with a sense of accomplishment and considered the 1987 protocols an environmental victory. We failed to grasp at the time how what seemed like simple and desirable changes in the operating structures for the agreement administration would fundamentally change the stature and independent voice of the IJC and its advisory boards. Years later, we would come to see how these changes would weaken the authority and independence of the agreement institutions in ways that even the proponents of those changes might not have anticipated.

The new 1987 Water Quality Agreement Protocols called for a direct exchange of information by the environmental agencies of each nation through something called the Binational Executive Committee (BEC), which eventual morphed into the Great Lakes Executive Committee (GLEC). Deliberations on cross-border strategies were no longer initiated through the "lake-centric" neutrality of the various IJC boards. Depending on who you talk to, the intent was either to improve accountability of the two governments or, if you are more of a skeptic, to bypass and minimize the independent voice of the IJC.

At their best, between 1972 and 1987 the Water Quality Board and Science Advisory Board had shined a light on the toxic crisis and pushed the cutting edge of scientific investigation without regard to how the findings might reflect on either nation. Their credible and sometimes outspoken voices had been one of the most potent tools of the agreement. Under the new arrangement, the U.S. and Canadian regulatory agencies were still deliberating the fate of the Great Lakes, but increasingly, through the lens of their individual national and agency interests, as opposed to a shared focus on the vitality of the Great Lakes as their reason for being at the table. The binational approach began to morph into a bilateral "my nation and your nation" approach, and we witnessed the fading of what had been a powerful early warning system for the Great Lakes ecosystem and a vibrant public constituency. It was one of the unforeseen losses for the agreement and the lakes. Still, the groundwork in the 1987 protocols helped drive meaningful change in toxic pollution cleanup and prevention in the Great Lakes region, in federal policy, and in global strategies.

CHAPTER TWELVE

Two Steps Forward

Building on the compelling scientific knowledge that shaped the 1987 agreement, we sought to drive progress in domestic law. Because so many toxic pollutants were airborne, updates to the U.S. Clean Air Act presented a rare opportunity for significant advances. The 101st Congress was inching forward on updates to the original law. Congressman Waxman and his ace team led the House strategy, and Cosmos caught their attention in 1989. We worked with Waxman's staff to craft language to specifically address airborne sources of Great Lakes contaminants, as well as coastal waters including Chesapeake Bay, and Lake Champlain (to win support from other delegations). The chemical target list is built upon the Great Lakes Critical Pollutants list and the 1987 agreement.

In 1990, when Congress adopted a comprehensive update, the new Clean Air Act included an expanded section on hazardous air pollutants: Title III. Section 112. Among its many subsections is a lengthy list of airborne pollutants, which, because of our efforts, included the most serious airborne Great Lakes contaminants. However, mercury, while on the list, received special treatment for its primary source, which I'll get to shortly.

Section 112 included a subsection (m): "Atmospheric deposition to Great Lakes and coastal waters." It acknowledged the connection between atmospheric pollution and contamination of the Great Lakes food web—something we had been seeking since our first Washington Week. It required a biennial report to Congress and five key actions:

1. Monitoring the specified waters for these substances
2. Investigating the chemical sources and deposition rates
3. Research to improve monitoring and to determine the relative contribution of atmospheric pollutants to total pollution loadings
4. Evaluating any adverse effects to public health or the environment
5. Sampling for the listed pollutants in biota, fish, and wildlife and characterizing the sources

Securing this language was one of my proudest moments in Great Lakes advocacy, but the glow didn't last. EPA could have done much more to leverage these provisions over the years, but didn't. In some administrations, there was little interest in doing so, and in others, inadequate resources limited good intentions. Sometimes this victory seems more symbolic than consequential. On the other hand, we have made significant gains in reducing these pollutants in the lakes, and along the way, EPA had to surmount many of its own hurdles, including the special carve-outs for mercury that ended up in Section 112.

Our little section survived the long and winding path that produced the 1990 Clean Air Act, itself a momentous undertaking. The act made new gains on old issues, such as reducing urban smog, but it also took on chlorofluorocarbons (a threat to the ozone layer), as well as fuel efficiency in cars, hazardous air pollutants, pollution from offshore oil rigs, reductions in carbon dioxide emissions, and acid rain. While the Reagan administration had avoided making commitments on acid rain, the Bush administration embraced market-based approaches (enabling sources to "trade" emissions under an established national "cap"). Thus in the 101st Congress, the Senate bill included acid rain controls along with the many other provisions.

As noted earlier, the acid in acid rain comes primarily from coal combustion. With Bush onboard, SO_2 and NO_x controls were coming, and the coal industry and coal-fueled power plants were, for lack of a better term, freaking out. They were also freaking out because coal (and oil) combustion is one of the primary sources of mercury in the atmosphere, and mercury was high on the list of chemicals the environmental advocates wanted to regulate.

Early in this legislative journey, members of the Great Lakes congressional delegation who supported us on toxic pollution made it clear they wanted to fly under the radar of the auto industry lobbyists on clean air issues. They worked out a deal with Texas congressman Mickey Leland, head of the Congressional Black Caucus, to champion the new regulations in Section 112. Representing a central city district with a majority-minority constituency in Houston, Leland was working on environmental justice long before it became a term. Leland, tragically, died in an airplane crash with fourteen others on a humanitarian mission to provide aid to refugees in Ethiopia in August 1989, and thus we lost our strongest defender of aggressive new strategies for controlling airborne toxics just before the last round of congressional deliberations was locking in language.

By the end of 1989, the Senate Committee on Public Works and the Environment had framed and fleshed out their best shot, and in early 1990, most of the substance of what was likely to be able to pass both houses emerged as Senate Bill 1630. While Great Lakes advocates had focused on subsection (m) of the Hazardous Air Pollutants title, elsewhere, the electric power-sector lobby, already actively objecting to the burden of addressing acid rain, had prevailed in getting special treatment for hazardous emissions from coal and oil-fueled steam-generating power plants, detailed in the new subsection (n). The new language was in the final package, and it changed the course of mercury controls in the United States (and some other significant toxic emission controls) for decades to follow.

The provisions enabled the power sector to play two strategic cards often used to delay or weaken environmental laws. The first was to require EPA to conduct a study on health risks from the hazardous air pollutants from these power plants. This would buy the utilities time before EPA could impose regulations. Thus, EPA would have to once again prove that mercury and other hazardous chemical emissions released into the atmosphere by conventional power plants were indeed hazardous, even though mercury and mercury compounds were well-known bioaccumulative chemicals with neurotoxic effects at the time.

The second tactic was to create ambiguity in EPA authority by providing *conditional* authority for EPA to regulate hazardous pollutants. The language stated that if, after completing the studies, the administrator "found such regulation appropriate and necessary," they could impose regulations. However, challenging whether an administrator's action is indeed both "appropriate and necessary" can provide more delays, while enriching the legal profession.

The full Senate began consideration in early March 1990 and, 124 amendments later, approved the bill on April 4. The House passed its version on May 23, with only eleven amendments, and then it was off to the conference committee and the White House. Once Bush had signed the new act into law, EPA undertook the effort to jump through the required hoops of the studies, and eventually seized the window of opportunity to address mercury from these sources. To no one's surprise, the administrator determined that it was, indeed, appropriate and necessary to regulate hazardous air pollutants from coal and oil-fueled power plants, just like every other source. EPA then developed Mercury and Air Toxics Standards (MATS) to regulate emissions from coal and oil-fueled power plants that generate more than twenty-five megawatts of electricity.

In 2012—almost twenty-three years after the Clean Air Act became law—EPA issued its final rule to control MATS, with industry compliance required by 2016. Legal and political wrangling ensued, with predictable challenges to those words "appropriate and necessary" and other strategies to delay, modify, or suspend this rule. But, with the rule and standards in place, utilities began to invest in new control technologies. In May 2020, the Trump administration's EPA stated, "After primarily considering compliance costs relative to the HAP (Hazardous Air Pollutant) benefits of MATS, EPA is concluding that it is *not* 'appropriate and necessary' to regulate electric utility steam generating units under section 112 of the Clean Air Act (CAA)."[1] If you are a utility that has recently invested in expensive new control technologies to meet the 2012 standards, I'm not sure what you are supposed to do with that determination.

Elections make a difference. In February 2022, the Biden administration issued this announcement: "EPA has proposed to *reaffirm* the determination that it *is* appropriate and necessary to regulate hazardous air pollutants (HAP), including mercury, from power plants after considering cost."[2] It is unclear how long it will take to reinstitute this program and how many new challenges it will face from opponents, but at least the wheels are in motion, and the standards themselves remain in place. It reminded me of Canadian environmentalist David Suzuki's version of the old saw about the nature of environmental progress: "In the environmental movement . . . our victories always seem to be temporary, and we keep fighting them over and over again."

Over the course of the delays, some states pursued their own regulatory programs for mercury, and rising concerns about climate change have driven major shifts away from coal and oil-fired power generation. EPA's website states:

> MATS, along with significant changes in the power sector, has achieved significant health and environmental benefits by reducing a broad range of hazardous air pollutants. By 2017 mercury emissions had dropped by 86 percent—down to approximately four tons. Acid gas hazardous air pollutants [e.g., SO_2 and NO_x that cause acid rain] and non-mercury metals [e.g., lead and cadmium, arsenic] are down 96 percent and 81 percent respectively when compared to 2010 levels.[3]

The EPA update is encouraging. Still, today, mercury remains a serious and extensive airborne contaminant in the Great Lakes, and other aquatic and marine ecosystems on earth. Globally, humans are still pumping 2,223,594 kilograms of mercury into the atmosphere each year.[4] I was surprised to learn that the largest source is artisanal or small-scale mining—mostly for gold—a particularly challenging source to regulate.[5] The parties to the young (established in 2016) Minamata Convention on Mercury raised this issue at their 2021 conference. Nurbaya Bakar, Indonesia's Minister of Environment and Forestry, pointed out "an alarming increase" in illegal trade in mercury.[6] Much of this is for use in the artisanal and small-scale gold mining sector. But the second largest source is still coal combustion at stationary sources like power plants.

More than thirty years after we pressed for regulation to protect aquatic ecosystems from mercury emissions, the Great Lakes states still publish detailed guides on what is "safe" to eat, and what is on the "do not eat" list for the Great Lakes and inland waters. EPA provides national guidance. A 2022 study said that only around half of the Great Lakes public is aware that these advisories exist, and more than five million people ate more fish than recommended by EPA.[7] Mercury is by far the most widespread contaminant, but PCBs are still prominent, and toxaphene and dioxin compounds are still on some lists. We lost decades of potential progress on mercury with one "small" compromise in subsection (n).

Would earlier regulations have been costly to the power sector? Almost certainly. Instead, we spread the costs to our biosphere, the aquatic food web, and our health. This is harder to quantify on a balance sheet than the cost and amortization of new control equipment and process changes, but costly nonetheless.

In addition to the CAA, we also made gains in the CWA. One bold move was a new section (Section 118) called the Great Lakes Critical Programs Act, which, among other things, required EPA to comply with several key actions embodied in the 1987 GLWQA—finally a way to drive these promises. When Congress passed the Great Lakes Critical Programs Act in late 1990, we considered it another triumph, but some of our aspirations were constrained by agency inertia down the road.

We hadn't anticipated that the tremendous legislative gains we made in the 1980s would stall out in the 1990s (and beyond) in agencies severely weakened by the budget and staffing cuts imposed by Reagan (and largely sustained by George H. W. Bush after his 1988 election). When George Bush campaigned, he claimed he wanted to be "the environmental president." At a major Sierra Club assembly in 1989, David Brower delivered a speech where he urged attendees to "thank Bush for saying he was an environmentalist, and then thank him again if he ever became one." The Bush administration was less openly hostile to the environmental community than the Reagan administration, but Bush's reliance on market-based approaches to solutions limited the horizon for other environmental solutions.

Still, the Great Lakes Critical Program Act compelled EPA to develop consistent water quality guidance across the eight Great Lakes states for the first time, and in 1995, they issued guidance for twenty-nine pollutants for all state and tribal regulatory agencies in the U.S. watershed, as well as anti-degradation policies. The Critical Programs Act also required Remedial Action Plans for U.S. AOCs (as well as those under shared jurisdiction with Canada), a Lakewide Management Plan for Lake Michigan, and studies, demonstration projects, and reports related to water pollutants and human and wildlife health.

These provisions empowered the environmental community to leverage administrative and legal action to drive progress, including pressing for alternatives to dilution-based approaches for toxic chemicals. While those actions fell short of achieving the zero-discharge and precautionary aspirations of the advocates, it moved the needle forward through other available channels, including new EPA controls for dioxin compounds.

The 1987 Water Quality Agreement, the CAA Amendments on air toxics, and the Great Lakes Critical Programs Act were my personal high-water marks for what the Great Lakes community accomplished during my years as Sierra Club's Great Lakes program director. These accomplishments were echoed by the landscape protections in the Michigan Wilderness Act, and national lakeshores. Behind the scenes we also secured substantial funding for EPA's and the National Oceanic and Atmospheric Administration's Great Lakes research laboratories. Superfund also helped advance toxic site cleanups.

In 1989, I enrolled in graduate school, working half-time on Great Lakes and other Sierra Club issues, and going to school full-time in what I remember as a two-and-a-half-year blur with limited sleep. I aligned my interests in the Great Lakes with my interests in communications and studied Great Lakes television documentaries for my master's thesis. When I finished in 1991, I was eager to ramp up my Great Lakes work again—so much so that I proposed to my bosses in San Francisco that I shift to full-time work on Great Lakes issues to really develop the program.

The option they offered me was that I could do this if I raised my own salary and stepped down from the senior position in the office. It was a reality check after ten intense years of helping build a new Great Lakes movement, and it required me to realize what the Great Lakes work meant (or didn't) to the organization. It was a tiny program in a large organization with broader priorities, whose heart was in the mountains. I could see that the ship was unlikely to turn in the direction I wanted. With anguish, I decided to move on, knowing that if I took the available option, I would always be in a peripheral role. I said goodbye to the field staff and the Great Lakes Program in May 1991. It felt like a divorce, and it hurt, but I knew I would stay connected with the issues and efforts to protect the lakes through other pathways.

Our intense effort on toxic pollution had ripples in wider circles. Our push to eliminate persistent toxic chemicals from the Great Lakes—especially the chlorine-based compounds that can generate dioxin—provoked pushback from the chemical industry. They organized in their own fashion, and began to play a much larger role in the IJC biennial meetings and other Great Lakes forums. This helped further hasten the retreat of the IJC and the federal agencies from big public meetings, and also from preventative strategies for toxic pollution. Meanwhile, the BEC moved ahead without the obligation to confer with the IJC, and IJC reviews and reports became less central to the deliberations between the two nations' implementing agencies. With fewer perceived threats to industry under the new GLWQA regime, industry resistance faded, as did their interest in the agreement.

However, the toxic genie wouldn't go back in the bottle, and the Great Lakes work served as a jumping off place for wider global attention on persistent toxic chemicals. Our early efforts helped drive the U.S. and Canadian production bans for many of the worst persistent and bioaccumulative pollutants. The pioneering work of the scientists we had listened to was getting wider notice and respect. Dr. Theo Colborn and her circle of colleagues were building a body of knowledge documenting the damaging effects of toxic chemicals in wildlife, humans, and ecosystems in the Great Lakes and far beyond. They were shedding light on the chemicals and compounds that could influence or disrupt hormone activity and framed the concept of "endocrine disruptors" as a group of chemicals that needed attention.

Beyond the bans, environmental groups were also pushing for industrial process changes in the 1990s, including eliminating elemental chlorine in manufacturing processes to reduce dioxin and furan compounds in the waste stream. The strategy was to eliminate some of the truly nasty chemicals and compounds at the source—embracing the zero-discharge approach we had championed a decade earlier. At the 1992 United Nations Earth Summit in Rio, persistent organic pollutants (POPs) gained international attention. This opened the door to establish an international treaty on persistent organic pollutants, and in 2001 the United Nations adopted the Stockholm Convention (the formal name for the POPs treaty) to address persistent organic chemicals in the global environment.

There has been tangible progress. Today's paper industry uses a lot less chlorine and chlorine-based compounds than in the 1980s, and that means fewer dioxin and furan compounds in the environment. Minnesota scientist Deborah Swackhamer was among the first to flag a correlation between chlorine used in the papermaking process and dioxin and furan compounds in their waste stream. The industry backlash rose to the level of attempts to cut her research funding, but she persisted. When you see a ream of paper that says "elemental chlorine free," you can remember Dr. Swackhamer, Henrietta on the buffet table, and the hundreds of other actions that helped move the global industry in this direction.

Throughout the Great Lakes region, while progress remains painfully slow, the United States and Canada have chipped away on cleaning up the forty-three Areas of Concern first identified by the SAB. The work is extraordinarily expensive, technically challenging, and imperfect, and presents unique challenges at every site. Some sites, such as Wisconsin's Fox River, have removed contaminated sediments through sophisticated vacuum-like dredging and treated the sucked-up sediments in facilities on land.

Other sites, like Ontario's Randle Reef in Hamilton Harbor, have entombed the highly toxic contaminants on site in clay, concrete, or impervious liners to isolate them from the environment. These strike me as permanent "sacrifice areas" in the larger ecosystem—a tragic but perhaps necessary choice in these times. Thirty-five years after the 1987 agreement, only nine AOC sites have made it to the stage called "delisting" (three in Canada; six in the United States), which means conditions will support a list of beneficial uses (which may or may not include safe fish consumption), and this allows them to step off the AOC list. We only have thirty-four to go.

Having ridden the crest of the second big wave of Great Lakes action, many of us from that era moved on to other jobs, issues, and changes in our careers.

The second wave was not enough. We still have a long way to go to restore and protect the chemical, physical, and biological integrity of the Great Lakes, and sometimes nature doesn't wait for democracy to get its act together before it's too late. We clearly need sequential, if not perpetual, strong waves of leadership and action as the lakes continue to face threats, old and new.

PART FOUR

Headwaters and Heart Places

There is a pleasure in the pathless woods,
There is a rapture on the lonely shore,
There is society where none intrudes,
By the deep Sea, and music in its roar:
I love not Man the less, but Nature more,
From these our interviews, in which I steal
From all I may be, or have been before,
To mingle with the Universe, and feel
What I can ne'er express, yet cannot all conceal.

—From *Childe Harold's Pilgrimage*,
George Gordon (Lord Byron)

CHAPTER THIRTEEN

The Imagined North

We cannot work at an intense pace all the time and remain healthy and sane. I found a scrap of notes from my goodbye remarks when I left Michigan that closed with a few observations. One of them was "Work on your passion until it hurts, and then go for a hike." Edward Abbey said it more elegantly when he counseled, "Do not burn yourselves out. Be as I am—a reluctant enthusiast—. . . a part-time crusader, a halfhearted fanatic. Save the other half of yourselves and your lives for pleasure and adventure. It is not enough to fight for the land; it is even more important to enjoy it. While you can. While it's still here."[1]

Many of us have "heart places" on the land or waters that call us back again and again, and many of mine are in the place called "up north." I am privileged to know them and hope the gift of what they give to me is reciprocated by what I have done to try to protect them from harm.

In the Great Lakes region, most of our major urban centers in the United States are "below" the 45th parallel on the charted globe.

Somewhere around or above that line is where the lands and waters known as "up north" begin to appear. The dominant landscape changes from farm and field to forest. Other than a few gashes of highway, the roads are two-laned. Towns become less frequent. It feels different.

In the urban imagination, the forests beyond the road are vast and wild, teeming with wildlife, rushing rivers, and lakes full of fish. Adventure, rest, clean air, privacy, and a good time lie at the "up north" destination. The imaged north is part truth and part fiction, and folks who live there know this. Many of the vast forests are the commercial timber production lands of international corporations, or public forests heavily managed in even-aged stands of red pine or aspen destined to become the next wave of toilet paper, disposable diapers, and office paper. The timber industry coined the term "beauty strip" to describe the row of trees left between highways and clear-cuts so that travelers wouldn't see the glaring slash of a recently mowed forest.

While the north woods has extensive forests, the "big woods" are a lot different from the one Laura Ingalls Wilder wrote about from her early memories of the 1860s. Even then, the logging boom was in full swing, transforming the ancient pine and hemlock forests, and then the maple-beech-birch forests into stump fields until the Great Lakes forests were "logged out" in the 1930s. State and federal efforts to reforest the landscape focused on fast-growing trees that had high commercial value, such as aspen and red pine. The forest economy shifted from saw timber to pulp and paper production. There was no concerted effort to bring back the original white pine and hemlock or mixed hardwood forests, in part because it is hard to get a quick return on investment from a tree that may take a century or more to mature.

The ancient forests of the Great Lakes region remain only in isolated pockets that escaped the logging boom by fluke or by the intent of private or Indigenous landowners. The big quiet of the big woods is isolated too. Within the patchwork forest, more and more lakes are ringed with cottages, and increasingly, large summer homes.

Many lakes buzz in the summer with large motorboats and jet skis. Rarely are you more than a few miles from the nearest road.

But still, there are places where the ancient trees were spared from the woodsman's axe, where the loons call on quiet lakes, where moose tromp through swamps munching lilies, where eagles soar, and a chorus of frog voices rises at dusk. This is the "up north" we imagine, and if we look for it, we can still find the remnants. If we work to protect and conserve those remnants and restore others, they will be there for future generations.

One of my favorite up north places is Sylvania, a wilderness of woods and water near the Michigan-Wisconsin border. The exact origins of the name choice aren't clear, beyond its appropriate connotation of a pleasant, wooded area, but I wonder if the men who first formed the Sylvania Club knew about Thomas Jefferson's old map and the proposed state of Sylvania, which encompassed most of what is now the U.P., northern Wisconsin, and northern Minnesota. If so, this is a fitting representative of what might have been the landscape in the early days of the republic, and they named it well.

In the out-of-print *A Sierra Club Naturalist's Guide to the North Woods of Michigan, Wisconsin, and Minnesota*, my friends Glenda Daniel and Jerry Sullivan offered this description:

> There is . . . a forest west of Watersmeet in Michigan's Upper Peninsula where sugar maples tower a hundred feet above the forest floor. White pines are scattered among the maples on high ground, while on slopes and in ravines there are hemlocks and ancient yellow birches, the bark of the latter so old that it peels in great sheets.[2]

Founded by a Wisconsin timber baron, A. D. Johnson, the Sylvania Club was originally composed of a small group of Chicago-based U.S. Steel executives who did business with the mining industry in Michigan's U.P. They purchased large tracts of land in the area for a summer retreat and private fishing and hunting reserve beginning in the late nineteenth century. They built spacious log lodges, staff cabins, and a clubhouse to accommodate members and guests, accessed by a network of primitive roads. The irony of timber and steel barons preserving a large natural area for their own enjoyment while extracting their capital from other lands is not lost on me. At Sylvania and many other areas in the eastern United States, large private holdings were one of the few ways that ancient forests were spared during the logging boom.

Over the twentieth century, as the original owners moved on or passed away, new members bought large parcels. One of the "second-generation" owners was Detroit's Fred Fisher, whose name was stamped on every car I'd known as a child in the form of "Body by Fisher." He died only a year after his purchase, and by midcentury his brother Lawrence and the Christiansen family of Wisconsin became the two primary owners.

After Lawrence Fisher and Clarence Christiansen died, the heirs faced the question of what came next. The Christiansen family enjoyed and used their holdings, but the Fisher heirs, aware of their father's interest in preserving the natural character of the place, favored a sale to the Forest Service. After Congress approved federal appropriations under the 1965 Land and Water Conservation Act, the Forest Service made a bid on the Fisher lands, and when the Christiansen's resources couldn't match the federal bid, they sold their lands as well. In 1966, we, the people, purchased eighteen thousand acres of land and lakes that Jefferson's surveyors had made available for speculators almost two hundred years earlier. Generations later, the largesse of the Fisher and Christiansen descendants to accept a public transition for this land gives me hope in humanity.

As the new owners, the Forest Service began to make plans to manage Sylvania as a public recreation area. They accepted bids to dismantle and remove the aging log and frame buildings, returning developed sites to natural areas. However, their early plans also included two thousand campsites on blacktopped roads. Given the exceptional water quality and fragile soils in the area, all that pavement and the soil compaction from thousands of carloads, their camping gear, and their campers would have fundamentally changed the character of the area. The dripping oils and greases from all those cars, not to mention sanitation facilities to accommodate thousands of campers, would have introduced pollution to some of the cleanest and least disturbed lakes in the region.

Many—from concerned citizens to Forest Service employees—argued for a different management plan, one that would protect the natural character of the forest and lakes. The modified plan included a small car campground, beach, picnic area, backcountry trails, and campsites, and set aside the vast expanse of lakes and forests for primitive recreation. In 1967, the Forest Service dedicated the newly named Sylvania Recreation Area with considerable fanfare, complete with a ceremony led by First Lady Ladybird Johnson. I like to think that she, as a great champion of wildflowers and natural beauty, enjoyed the view of the meadows and Clark Lake that September day.

My brother ventured to Sylvania with some high school friends in the early 1970s. He returned with stories about towering trees, how clear the water was, and how far down you could see below the surface. He also described some very hardscrabble camping without modern campground conveniences, which meant he would not return.

I was delighted to discover the area firsthand through the Sierra Club's work on the Michigan Wilderness Bill. After our 1978 fact-finding and photo blitz, I returned to Sylvania the following year with the new Sierra Club chapter chair, Ken Magnell, to show him what was at stake. We had the luxury of several days and his canoe to explore beyond Clark Lake. On the first day, we were enjoying lunch at a picnic table under the shade of ancient maples at one of the backcountry campsites when we heard the angry buzz of a chain saw rising in the adjacent campground. We walked over to investigate and had a philosophical discussion with the Forest Service employee about whether tree trimming was appropriate in proposed Wilderness areas. (We were objecting, also, to the noise, having come a considerable distance to sample the peace and quiet.)

He explained that they were trying to remove trees that posed a hazard to campers—especially towering old trees close to tent sites. In a friendly touché, he then raised the point about whether human-made picnic tables belonged in Wilderness. He returned to his chain saw and we to the picnic table to ponder this further. Years later, both the chain saws and the picnic tables were gone from the interior of the Sylvania tract, but these are the sorts of nuances that Wilderness management agencies and their publics struggle with to maintain wild places. Often, the public's expectations are different and sometimes in conflict with the challenges faced by the agency and the needs of nature herself.

I returned again for the 1984 Senate Wilderness tour, accompanied by staff from Michigan senators Levin and Riegle's offices, some key committee aides, and assorted Forest Service staff. Setting out from the Clark Lake boat landing, our Forest Service hosts paddled this esteemed group to Glimmerglass Lake, a sheltered lake attached to Clark Lake that lives up to its name when the air is still. It was showing off that day, and we admired the reflections of towering pines in the mirrored surface of the water.

On our trip back to the boat landing, we headed into a rising wind. (Why is the wind always in your face on the way back to your campsite or boat landing?) Levin's Washington aide was bobbing in the bow of one of the canoes that had a strangely mounted seat that seemed too high for a safe center of gravity. A Forest Service staffer handled the stern, pushing against the waves, while our important passenger held on in the bow without the skills to help brace the boat through the breaking waves. I thought we might lose the senator's fragile support at any time if they capsized. I breathed a sigh of relief when they safely reached the boat landing on the far shore.

While Sylvania's twenty-one thousand Wilderness acres are largely intact and have no remaining homes, roads, or motorboats on them, there are a few exceptions. Crooked Lake, which reaches into the middle of the Sylvania tract, had homes, cottages, and a few resorts on its north bay before the legislation was even in the idea stage. While Wilderness designation appealed to me and many others who wanted to see the wild character of this rare place protected, it did not appeal to people who wanted to continue to use their motorboats on Crooked Lake.

When Reagan signed the hard-fought little Michigan Wilderness Act on December 8, 1987, nineteen thousand acres of the Sylvania tract (excluding Crooked Lake's north bay and its homes, cottages, and resorts) were included among Michigan's ten new official federal Wilderness areas, which together totaled 91,535 acres—less than three-tenths of a percent of the land base in Michigan—not a major shift in Michigan's land management, unless that small percentage affected you personally. The controversy of what was or wasn't allowed on Crooked Lake remained.

After the Wilderness designation, the Forest Service issued a decision in 1996 to allow only small electric motors within the wilderness portion of Crooked Lake as well as Big Bateau and Devils Lakes on which there also had been preexisting use of gas motors. Two Crooked Lake property owners (a resort owner and one other) sued over this decision, and federal courts ruled that the two plaintiffs and their guests are permitted to use gas motors on the entire surface of Crooked Lake. The resort has since been acquired by an owner who doesn't allow gas motors, so only one property owner and their guests still have this privilege.[3]

Motors still buzz along on the northern bay of Crooked Lake today, and you can hear them at dusk throughout the Wilderness area when the wind has died. But thirty other Sylvania lakes are the domain of canoes, kayaks, and loons. We owe the public access to these quiet lakes and forests to the thoughtful conservation of the original inhabitants in the region; the self-interest and then the largesse of a collection of twentieth-century steel, timber, and auto magnates and their heirs, coupled with immeasurable hours of volunteer efforts to pass the Wilderness bill; the stewardship of the U.S. Forest Service and volunteer organizations; and the affirmative vote of the United States Congress.

The Michigan Wilderness bill was a hard pull on steep ground. After the legislation finally passed, I had no desire to go back to places where there were fresh wounds on all sides. I didn't want to reexperience all the raw edges. As wise Sierra Club colleague Tim Mahoney counseled over strategy lunches at the Tune Inn on Capitol Hill, "Never vacation in an area that isn't saved yet."

I did recommend it as a weekend destination for my husband, Bill, and our botanist friend Marty. Given Marty's botanical interests, he wanted to check out a place known for its rich array of native plants, and off they headed for a summer's weekend. They were having a fine time camping at one of the designated campsites on Loon Lake. Then, a ripsnorter of a summer thunderstorm blew through in the middle of the night and snapped off the top of one of those huge old maple trees—just like the one my colleague and I had defended from the chain saw years before. It landed on top of their tent.

In what was surely some sort of divine intervention, a V-shaped intersection of a major limb and the trunk landed in such a way that it engulfed but did not crush them. The tent poles were twisted like pipe cleaners. As their story goes, after the initial shock, and ascertaining that they were both unharmed, they turned their attention to staying dry in the collapsed tent as the storm continued. They determined that they would sort it all out in the morning and went back to sleep.

Most of the designated campsites at Sylvania come in pairs. Two sites, accommodating two different parties, are located close together, maximizing good landing sites on the lakes and minimizing the human footprint in the backcountry. The next morning, the adjacent campers came by to see how their neighbors had weathered the storm and, on seeing the tent, thought they would have to haul bodies out from under the downed crown of the maple. Instead, Marty, awakened by their concerned conversation, managed to poke a head out one end of the remains of the tent and affirm that they were alive and quite well but really bummed about the tent. Marty is the sort of fellow who doesn't let a near-death experience get in the way of an otherwise pleasant morning. Bill later referred to the incident as "the tree that bit my tent."

Perhaps that incident was a little too vivid and the legislative conflict a little too fresh for me, but Sylvania wasn't on the short list for future vacations, and we set our sights on farther destinations on our "at-least-once-in-a-lifetime" list. Over the next fifteen years, that took us back to the Rockies but also to the Arctic Circle, the Boundary Waters, the Serengeti, Kauai, and other memorable destinations.

When we chose to become parents, we knew our adventures would take a different direction. Our son Colin arrived in November 1994, and the next summer, I started a new job heading up the

Biodiversity Project—an initiative by a consortium of foundations to advance awareness of biological diversity among U.S. voters. The consortium pursued this in part because the United States (joined only by Andorra, Iraq, and Somalia) had declined to sign on to the United Nations Convention on Biological Diversity in 1993, largely due to Senator Jesse Helms's opposition.

I was a new parent in a new job. While we have friends who will enthusiastically plunge into wilderness adventures with a baby in the backpack, it requires extraordinary planning skills along with creative strategies for the diaper thing. Between the demands of my job and a little one who considered sleep an avoidable nuisance, I was so sleep deprived in the early "mom" years that all the extra effort seemed too daunting to consider. As he got a little older, we started out the simple way, with a few car-camping trips to see how it all came together with a new partner and his gear in the mix.

Car camping was OK for Colin's introduction to tents, sleeping bags, and cook stoves, but at the end of the day, you are essentially in a parking lot with lots of neighbors, even if it is a lovely parking lot. We considered how to take the next step to a wilder and more remote experience. We loved the Boundary Waters, but for a child who was not thrilled about walking long distances and not able to shoulder his own share of the gear, this seemed like a plan for daily negotiations in "have to" territory, as in, we have to carry our gear over this portage, you have to keep up, we have to keep going until we reach our campsite (whether you want to or not). We enjoyed a week in a little Isle Royale rental cabin while Bill's folks stayed at the lodge. They watched Colin during the day while we tromped around the west end trails, paddled in Tobin Harbor, and poked around Rock Harbor in a small motorboat. This was great fun, but expensive and not easy to pull off without considerable planning, travel time, and helpful grandparents.

Eventually, we considered Sylvania again. It was a short day's drive from home, with backcountry campsites still relatively close to the car in case something went wrong, and, with careful planning, easy portages. We were beyond the preschool years and at a point where, at age six, Colin had enough smarts about water and safety to give us confidence that he'd follow instructions and be successful as a happy camper.

In 2001, we registered for a permit and dusted off the canoe gear again. We pointed the car north and were on our way. Things began to look familiar as we turned westward on Highway 2 in Watersmeet. We stopped to pick up a few things at Sylvania Outfitters, because we like to make a practice of patronizing people like its proprietor, Bob Zelinski, who supported the Wilderness bill in the tough years. Then we were back on Highway 2, looking for the road that leads into the tract, and there it was at the top of that rise. We stopped in at the A-frame hut that served as the registration site for Sylvania camping permits in those days. We confirmed our campsite and went over the rules, which hadn't changed much over two decades. Then we headed down to the boat landing on Clark Lake. Our energy started to rise with the anticipation of spending time with an old friend again.

We unloaded the car, packed up the boat, and were eager to get started. We filled the water bottles at the boat-landing pump with the cold, clear, mineral-tasting water of the north, did a final gear check, slicked on the sunscreen, and then we were ready. We seated a slightly anxious Colin in what we call the "Queen of the Nile" folding seat in the middle of the boat and held it steady. I climbed in the bow and braced with the paddle while Bill climbed into the stern, and then we were off, gently bobbing in the waters of Sylvania again after almost twenty years. A few strokes later, the boat landing and parking lot began to slip away behind us along with our city selves. There again was the little island with craggy spruce and pines off the bow, and off to port, down the lake, the arm of land that marks the narrows near the Balsam campsite.

I didn't expect the little tears of joy that welled in my eyes as the loons by the island called out a "welcome back." All those years, all those hearings, all those trips to Washington were worth it.

In Wilderness, more than any other place I know, you can go home again.

CHAPTER FOURTEEN

Refugia

The Great Lakes region has many subcontinental divides that determine whether the waters will flow into the Mississippi Basin, Hudson's Bay, or the Great Lakes themselves, and onward to the Atlantic Ocean through the St. Lawrence River. Sylvania is such a place, and from its sparkling lakes the waters flow to either Lake Michigan or Superior, or the Wisconsin River watershed, which is part of the Upper Mississippi Basin.

The nearest town to Sylvania, Watersmeet, is aptly named, for it is here where the waters of major watersheds meet and part. There is no dramatic statement in the landscape that points this out. You can, however, feel it ever so gently in your knees when you make it to the top of the portage between Clark Lake and Loon Lake and begin the descent to the other side. This portage takes you over the Lake Superior/Mississippi River divide, but a maple tree on one side looks just like a maple on the other because you are at the very cusp of two watershed ecosystems whose distinctive characters will be shaped downstream.

Sylvania is fed only by rainfall and snowmelt, and its waters are elegantly clear and clean. There may be some nutrient infiltration from the handful of nearby septic systems or fuel and oil dripped from boat motors or cars parked at boat landings. And unfortunately, wherever people go, there might be some inconsiderate waste left behind by sloppy campers. So far, these provide few visible clues of their presence in the waters of Sylvania, which offer exquisite pleasures for humans and outstanding conditions for wildlife that require clean water.

I live in a city where the nearshore waters of our nutrient-rich lakes are often a blooming chartreuse soup by late July each year. Ongoing cleanup efforts will take decades to effectively reduce the nutrient pollution that fertilizes the algal blooms in Madison's lakes. As a result of living so close to "green water" lakes, we are always amazed to set out on these cool, crystalline waters under summer skies. When the winds are calm, you can float over big rocks, sandbars, and an ancient log or two, visible at depths that always take a little mental recalibration to process because they are so removed from what would be "normal" in so many other places.

The sandbars of Clark Lake are among my favorite swimming places. They change every year with currents and water levels, so one has to be open to a new experience each trip, as during high-water years, they can disappear below the water. The clear water over submerged golden sands—sometimes a foot or two deep near shore—warms up on a bright summer's day, allowing even a wimp like me to enjoy a swim. For the more adventurous, the deeper, dark, cool waters are just a step off the edge into the abyss (and a good reason to be careful if you aren't a strong swimmer). Swimming in water like this is such a luxury: no algae, no smell clinging to your skin or swimsuit, no tangle of weeds—just gorgeous.

After a swim, I love stretching out on a big rock or a stretch of sun-warmed sand. I drop down a few notches in evolution and reclaim my inner lizard, restoring my body heat in the old-fashioned way. Sublime.

Beyond human pleasures, the clear waters provide myriad other forms of life with exceptional habitat. My decade with the Biodiversity

Project sharpened my awareness of species and habitat and my appreciation for healthy, diverse systems. We always see eagles here, and it isn't just the occasional "oh wow" moment. After a day or so, we can identify the adults and juveniles, who is using familiar nesting sites, and where they are fishing this particular year. I never knew eagles could swim until I saw one do it by our campsite on Loon Lake. Colin and I were at the little sandy landing by our campsite, puttering around while Bill was out exploring other lakes. An eagle came swooping down, talons out, and latched into a large fish just below the water's surface.

The problem was that the fish was too large, and the bird couldn't raise itself and its catch out of the water. It struggled a bit, and I found myself foolishly thinking I should offer assistance (as though I would know what to do and that wouldn't put the eagle into greater panic). Instead, we watched from the shore as the huge bird used the shoulder of its wings and awkwardly paddled itself to the nearby opposite shore, dragging along its sea anchor of a fish. At that moment, we didn't know, but learned later, that once its talons are engaged in the "lock and load" position, an eagle cannot release its grip on a fish until landing, and that sometimes the birds miscalculate and don't make it to shore.

This eagle heaved itself up on the rocks, where it opened its wings to dry its wet feathers. It simply rested for a while, and we weren't sure if it was going to make it. After a bit, it fluffed its wings and began to tear into the fish with its beak. After a snack and another short rest, it lofted into the sky with half a fish in its talons and headed off, presumably to the family nest. Moments later, a second eagle—we assumed the mate—swooped in and made off with the rest of the prize. We were pleased for the happy ending (at least for the eagle). A show like this on Loon Lake is better than watching PBS.

Sylvania is high-quality eagle habitat for many reasons. The many ancient trees provide tall perches in which to build nests and scan the lakes for fish. The lack of disturbance from human settlements gives eagles a chance to just be eagles. Then there's that water. Clear water makes spotting a fish while on the wing all that much easier, and clean water means a healthier food supply of less-contaminated fish.

All this, and at twenty-thousand-plus acres, there's enough space for several pairs to make nests and feed themselves and enough other eagles around to find a partner when they reach breeding adulthood.

The eagles always command our attention, but many other wonder-worthy species thrive in the clean lakes and ancient forests. Old dying and downed trees are heaven for woodpeckers and other insect eaters. Dozens of species of songbirds live in the marshy areas, the sheltering branches of the hemlocks, and in the hardwood canopy. Bill, who is a serious birder, is often out scouting in the mornings with a battered old *Peterson Field Guide to Birds of North America* in hand, and rarely comes back disappointed. One afternoon in late August, we stood below a cluster of shoreline trees filled with warblers frenetically feeding on insects in the waning day, fueling up for the migratory journey ahead.

The water birds are among my favorites. We love the parade of a merganser flotilla paddling along the shoreline, dabbling and poking in the shallow reeds and stony edges. Mergansers subscribe to the cheaper-by-the-dozen theory when it comes to raising young; in summer, they typically have a big bunch of babies trailing along behind the leader. Apparently, this is a survival adaptation, with one seasoned mom playing day care supervisor for the offspring of many other birds. How they pick the mother of the year is a mystery. With the many birds of prey hunting from above and the threat of northern pike hunting from below, I suspect there is safety in numbers, along with some entertainment value for admiring humans.

Above the lakes, sometimes we'll also see ospreys, various hawks, and great blue herons. The star of the lakes is, of course, the loon.

From a design perspective, they are drop-dead gorgeous. They have sleek lines for diving and flying, and the spectacular black head punctuated by garnet red eyes rises above a body with black-and-white herringbone patterns in complex and perfect balance. Below the waterline, strong feet power them through their dives and in takeoffs and landings. As if being strikingly beautiful weren't enough, they are blessed with the voice of the north—the piercing melancholy music of lake country. There is the rise and fall of the minor-key voice that calls at night, the twilight lament that sounds like grief itself ("Where

ARE you?"). Then, we might hear the raucous laughing call of loons in flight—the source of the word "loony." In quiet pairs, they might offer a gentle hoot or two in what seems like a private conversation. For defense, they issue a loud "Away from here, away from here!" staccato protest of territory breached. In quiet lake country, it echoes across the water, fills the space, and rests in your bones.

The Forest Service has rules about how close you can get to loons in a canoe, and we're always happy to comply. However, sometimes the loons don't read the rule books. One afternoon, we found ourselves being shadowed by a mom and chick. We'd paddle away, and she would close the distance. After a bit of this game, we just drifted, and so did she, encouraging the little loon to dive as she did. Perhaps we were floating over a good fishing hole. Perhaps she just wanted to let the little guy know what humans and canoes were about. Eventually we paddled on, and so did she.

The loons of north country lakes are in trouble. Loons need big lakes for their awkward takeoffs—they need to work up a head of steam to loft their bodies into flight. They need clean water for the same hunting reasons that the eagles do. They also don't like disturbance—from humans, machines, dogs, predators . . . you name it. A tremendous amount of former loon habitat is disturbed in one way or another, and there are fewer and fewer places for loons to feed and breed. They are also extraordinary consumers of fish.

Remember those trace pollutants that come in from the atmosphere? In northern lake country, mercury is the big problem. Just as in the Great Lakes, the source is rain and snow washing mercury compounds out of the atmosphere. In smaller inland lakes fed only by rainwater, there's no dilution capacity comparable to the enormous volume of water in the Great Lakes, and mercury will build up much faster in an inland lake food web.

The northern cold-water lakes, with their short food chains, are ideal concentration systems. Humans are warned not to eat too many fish from inland lakes—especially the larger predator fish. Loons can't read the warnings, nor do they have alternative choices in their diet as we do. Increasingly, mercury is thought to be a factor in declining loon populations, coupled with the impact of development (and thus habitat loss) on northern lakes and along their migratory corridors. Only a handful of loon voices can fill the night, fooling us to think they are widespread, but they are indeed few and far between.

I cannot imagine the north country without them. For now, in places like the Boundary Waters, Isle Royale, and the less developed lakes across the northern forests stretching into Canada, there are still loons. A hundred years from now, I hope they will still be here, instead of only questions about what loons were.

Other voices join in the night music of Sylvania, and they too are here because of the clean water. The place is hopping with all kinds of frogs. Amphibians are remarkable indicators of environmental quality. Through their skins, they literally absorb the environment in which they live. Throughout the world, frog species are in terrible decline, and extinctions among frogs and other amphibians are happening at a frightening pace. They are barometers of the world's environmental quality, and the readings don't look good.

In much of the Great Lakes region, you have to be a robust frog, like a bullfrog, to get along these days. Many other frogs that used to swell the night chorus have quietly disappeared from the landscape. As a result, we were delighted one morning to come across four different frog species on one little stretch of a Sylvania beach, out sunning themselves by the water's edge. We realized we had no frog smarts anymore, and other than identifying one that was clearly a leopard frog, we had no idea who or what we were looking at (presumably, neither did the frogs looking back at us). One was a tiny little rusty-brown fellow, another midsized and green, another blackish.

It made me realize that we'd lost a bit of knowledge about the world of critters, because it is no longer part of our normal experience. I wondered how many other nuances of the natural world are slipping away from human knowledge and familiarity because they are quietly fading from the fabric of our lives. I'm thrilled that the frogs are doing so well in Sylvania. I'm sad that it's one of the few places we hear their voices anymore.

When camping, we turn the canoe over at night in case it rains and to keep it from blowing around. One morning, we flipped the boat back over and discovered that we'd inadvertently crushed a little frog when we'd pulled it up on shore. Surely one of the carrion-eating birds would have it for lunch, but it left us feeling sad and responsible that with all the odds against them, our canoe was one more threat to an uncommon species' survival.

After that incident, a friend sent along a short essay by George Rabb, one of the world's great champions of amphibians, and one of the men responsible for changing the mission of zoos from cages to conservation in the twentieth century. The essay recalls a night he spent comparing frog choruses from one swampy area to another.

> As one comes nearer, different species and individuals cautiously drop out of the chorus so that it is no longer the glorious cacophony that originally attracted one's attention. Indeed, as one moves even closer, the entire chorus is likely to shut down. . . . When various members of an orchestra drop out, the range of music in the group's repertoire diminishes and future concerts will be correspondingly limited. In biological terms, when species disappear, ecosystem functions are impaired and evolutionary potentials vanish.[1]

Our world needs these voices. George helped drive international efforts to rescue species threatened by the worldwide fungal disease pandemic that has wiped out so many frogs (a significant but not sole cause of their demise). He also elevated the parallel need to protect the high-quality environments in which frogs still thrive—those are the front lines between amphibians and extinction. We are living through the sixth mass extinction (the other five having taken place prior to human civilization). We can blame asteroids, prehistoric volcanoes, or other ancient forces for the prior five mass extinctions. This one is caused by humans. In my work—especially at the Biodiversity Project—I was constantly dealing with the hard news of extinction and loss. Time spent in places like Sylvania gives me hope that there are still some refuges where nature is holding on and thriving.

Sylvania, though, is a tiny island of wilderness. Even though twenty thousand acres sounds big, it really isn't. Sound travels so easily over water. On quiet nights when we are sitting by the lake, you can hear the distant sound of cars and trucks shifting gears on the roads that border Sylvania. Sometimes at dusk, you can hear the Crooked Lake motorboats heading home for the day. We haven't quite managed to get away from "it all," and neither have the lakes or the life they support.

In 2005, a team of seasoned Great Lakes scientists authored a paper entitled "Prescription for the Great Lakes Ecosystem." They described an "ecological tipping point" past which ecological recovery would be increasingly difficult, if not impossible. They argued that one of the most important things we can do to restore and protect the Great Lakes is to clean up the ecosystem's "arteries"—the major river systems that flow into the lakes every day. It makes sense. If you keep pumping sewage, sediment, agricultural chemicals, and a dollop of industrial goo into the rivers that feed the lakes every day, the patient isn't going to get better. The big rivers are fed by many tributaries, which have their origins in headwaters. A few headwaters are like Sylvania—clear and remarkably healthy. But throughout the Great Lakes ecosystem, many other headwaters aren't anymore.

In the end, high-quality habitat for loons, eagles, and leopard frogs means high-quality habitat for us downstream: the cleaner the headwaters, the cleaner the lakes. We don't have to designate Wilderness areas to protect high-quality habitat, but it seems to be one of the few ways we've been able to count on that can really get the job done. Still, little islands of wilderness on the landscape are simply not enough to provide the habitat, clean water, and fresh air we all need, and they never will be.

We must design and embrace better strategies for all the places in between, from trout streams to backyards to city streets. As the fish and the loons know, what goes around comes around: the mercury from our power plants goes into their diets, and ours. There is no safe "away" place in which to hide the messes of our civilization or to escape from them.

The diversity of life and the quality of water in places we think of as wild and natural are not relics or museum pieces of a place set apart from civilization. They are, instead, living indicators of conditions we should strive for to keep our chorus, and that of George's frogs, singing in the night. All of the earth's water is ancient, and it's all we're going to get, regardless of which side of the divide the rain falls. Water isn't a commodity—it is life, and the quality of our lives depends on the quality and abundance of the earth's water ecosystems. Here in the freshwater heart of North America, we have no greater responsibility to humankind than to protect the vitality of waters that sustain life on Earth, now and always.

CHAPTER FIFTEEN

Mosquito Beach Jubilee

Once I had the courage to return to Sylvania, other reunions became possible. One of the places I had loved most was also the source of considerable anguish from the debates, compromises, and losses of the 1970s and 1980s. I hadn't been to Pictured Rocks since my years with the Sierra Club field staff in the late 1980s, but I longed to see special places there again. In August 2004, we rented a small, rustic cabin on Lake Superior just west of Au Train for a few weeks' vacation and invited my childhood friend Kathy and her husband, Mike, to join us for a few days. Our plans included a hike at Pictured Rocks.

So, on one of those rare August days when the elements come together—blue skies, cool breezes, no bugs—I found myself again on the lakeshore trail at Michigan's Pictured Rocks National Lakeshore along the familiar stretch from Miner's Beach to the Mosquito River. Our destination was Mosquito Beach—a name that predisposes people to dread and avoidance, but can serve as a deterrent to excessive traffic in a fragile place. The mosquitoes and black flies can be fierce here in June, but the same can be said for just about any place near ponded water (for mosquitoes) and running streams (for black flies) in the north woods. Then there's the combination of mosquitoes and biting stable flies on hot and humid days in July and sometimes even August. So, seasonal timing and pleasant weather are likely to determine whether you will remember a trip to "the Rocks" as a slice of heaven or bug-driven torment.

By late August and early September, nearly everything in the summer insect community has already hatched and moved through the life cycle stages where they seek the flesh and blood of mammals. The weather tends to be drier (fewer ponds and puddles for mosquitoes to incubate), and the lake has had a few months of solar radiation to warm up to a point where swimming might just be enjoyable. It doesn't mean you should leave the bug jackets or repellent behind, but the odds are better for more pleasure and less misery than, say, the first week in June. People who spend a lot of time in the outdoors in the north country know this window of opportunity, and so these few precious weeks of late summer tend to attract crowds of campers, canoeists, and anglers seeking that ideal triangulation among warmth, low bugs, and seclusion near big trees and water. Weather, of course, is more the luck of the draw. Sometimes the stars align, and this day was one of them.

It had been far too long since my last visit, and I was eager to be part of this place again. As I walked with my husband, son, and lifelong friends, I realized that I had shared time in this place with some of the people I most love on the planet. I had always known joy and beauty here. We were all glad to be here together, and the chemistry of reunion and anticipated discovery had us in fine spirits for the three-mile walk.

We started with the steep climb from the parking lot that always tests car-stiffened joints and reminds us, as we walk closely along the exposed sandstone wall, that much of this walk will be on an escarpment. We threaded along the edge of the "glacial rebound" where the compressed surface of the earth was freed from the weight of the ice sheets that once covered this landscape a mere ten thousand years ago. The land rose some two hundred feet during its rebound and is rising still, ever so slowly, even today. This is a young landscape in Earth time. In human time, my knees and lungs were telling me that I had a lot less mileage on the chassis the last time I climbed this hill, and I was grateful to be carrying a light load on this hike. We stopped to catch our breath at the top, and a cooling breeze rushed through the birches.

We struck out on the well-established trail that takes hikers inland for a while through the maple-birch-beech forest. As we walked, my memories stirred and my senses were rewarded. I felt lighthearted. The trail felt good underfoot. It winds around the marshy areas created by the lakeshore's vigorous beaver population (who have been held responsible for taking some of the thunder and show out of the waterfalls that tumble over the cliffs into the lake). Ah, nature's little Corps of Engineers, damming streams for their own species. After what the voyageurs and European fashion did to the beaver, I was willing to cut these cheeky rodents a little slack for turning this stream into habitat instead of scenery. Someday they will move on, and a spring flood will crash through their abandoned dam, and the waterfall will awe visitors again. For now, the stream belongs to the beavers.

Soon the trail brought us back to the escarpment's edge. We stopped at each little lookout—the spur trails are easy to recognize—and paused on each to enjoy the breeze off the lake. We listened to the sounds of waves crashing below on the arching walls. Our gaze took in the vistas from the cliff edges—blazing blue water; vertical sculpted walls of gray, amber, and russet sandstone; and the green forest. In a place like this, you run out of descriptive superlatives as you try to respond to the view from the lookouts. This palette of colors speaks to some inner sense of affirmed beauty in the human fiber. Somewhere in ancient memory, the senses say: "Abundant clean water and green forest; this is a good place."

The vertical drop was as impressive as I remembered, and the depth and clarity of the blue-green waters seem to call out to be enjoyed. In the quieter coves, you can hear the splashing and gulping sounds of waves captured by pockets and sea caves below. Part of the brain responds to this dramatic scene and says, "Don't you just want to dive into these cool, turquoise waters?" But another part says, "instant death, no way out," and it wisely overrides the tantalizing pull of the sparkling water far below.

Farther down the trail, we stopped for a little rest and water at the grassy place surrounded by birches that had served as a good spot for a break on trips long ago. If memory serves, this was once the site of a cabin long before the lakeshore was established. One can imagine someone looking at this view and choosing to set up housekeeping here, for it is such a pleasant place to be alive. On this day, we rested on a log and offered encouragement to our young hiker who was skeptical about all this walking and how long it was taking (and a little unnerved by those sheer drops we had shown him). We let Kathy and Mike strike out ahead so we could keep a slower pace with Colin.

About an hour and a half down the trail (with a nine-year-old, a speedy hike is not the goal), my heart lifted at the place where you can catch the first glimpse of the broken sandstone shelves that mark the beginning of the beach. The submerged slabs glowed in the sunlight through turquoise and chartreuse water. My anticipation rose as we drew closer to a treasured place held only in memory for so long.

First, there is the steep descent from the bluff, and then the smell of the balsam forest along the lower lakeshore path. My pace picked up as I looked around, affirming "It is still here, the way I remembered." I was flooded with joy and relief at this reunion. I felt welcomed home by the birch, balsam, and white pine. I breathed deeply, glancing all around me, sensing the magic in this place. I released the cautious defenses that I carry into so many places that I have once loved and seen degraded or destroyed. Gratitude washed over me. It was still whole and beautiful. I took in the sweep of the beach, the forest, and the river rushing into the lake. Jubilation.

We spilled onto the beach before the bridge, as we have done before, because it pulled us. We stopped to rest and eat on the rounded rocks by the mouth of the river where the cool waters emerge from the dark forest and tumble over hundreds of minute shelves of blackened sandstone to rush into the cold surf of Lake Superior. I was the definition of the happy camper.

Colin needed no further urging in this perfect spot. Shoes were off and pants exchanged for the swimsuit in the daypack. Then he plunged into the water. Within moments, he was gleefully being pummeled by the breaking waves and the mixing waters of lake and stream. I knew we would have to pull him out later, but that could wait. Soon our boots were off, too, and we were wading in the river, slipping on the stones, dampening rolled-up pants, and feeling deep-down contentment.

We picnicked on the rocks: gorp, cheese, and crackers for us, while Kathy and Mike opted for some power bars that look packed with protein and essential nutrients but lack the memories of a hundred hard hikes and portages fueled by Peggy Wayburn's gorp recipe. The origins of gorp were "good old raisins and peanuts," which have inspired many more exciting variations over the years. Peggy was half of the remarkable team of Peggy and Ed Wayburn, lifelong heroes of the wilderness movement, champions of wild Alaska, and each distinctive characters. I had the pleasure of knowing them during my Sierra Club years, and when *Sierra* magazine published Peggy's gorp recipe, it seemed worth a try. We've been munching her recommended mix of dried apricots, cashews, and M&Ms on the trail ever since, and it allows us to think fondly of the Wayburns on every trip.

Part of what enriches time spent in the wild is the collective memories it evokes. There are parts of our souls that are shaped by time in the presence of vivid nature, and remembering renews those gifts. We will tell stories in tents, on ledges, or around campfires, of eagles seen, great storms that pummeled the tent, hapless companions with excessive weight in their packs, moose surprised on the trail, the best night of stars ever, bear noises, and acts of great courage (or folly) in rapids, crossing streams, or rescuing injured friends.

These are rarely conversations of the living room or dining table at home. Perhaps they require a setting without a roof or cushioned furnishings. Maybe they need the yeast of tired muscles, a damp bottom perched on a log, and the smell of woods in the waning light of the day to rise gently from memory.

This being a day hike, our time was measured, and I wanted to explore farther afield and rekindle some memories of my own. I went through the ritual of attempting to brush off wet, sandy feet and put them back into dry socks, which is always a struggle. A bandanna helped disperse the grit, but I still had to tug to persuade the socks to go over reluctant feet. Socks on, I shoved the boots back on for a tour of the old campsite, while Bill kept a close eye on the happy kid in the water.

I scrambled up to the trail, crossed the wooden bridge that spans the Mosquito River, and had a look around the campground. Of course it had changed, but there were familiar signs. Too many people have loved this place, and the Park Service had moved the campground back into the woods away from the shore some time ago. I'm sure it was necessary, but I'm saddened that it is so. The sandstone bluff seemed a little worn and battered (so hard to resist the pull to scramble over the brink, to play on the ledges, to dangle feet on the fragile edge). Young trees were sprouting up, and lots of vegetation covered the places where the old campsites were. This tells me that there is hope that the forest will reclaim its place here.

I recalled my first solo trip and learning that while solitude has its virtues, sharing the beauty, the risk, and the weariness at the end of the day was something I needed and wanted. Over the years, I had come here many times with others, often for official reasons—trying to make the case for keeping this place and others in the Great Lakes as wild, natural, and road-free as possible.

This is where we introduced Dave Dempsey to the Rocks with the "Jane Elder Pictured Rocks immersion course," which included a dose of my dubious camp cooking. (For the record, even in a pinch, don't mix butterscotch and lemon pudding to save washing an extra pot.)

Over the years, Dave and I became friends as our lives and careers wove in and out of efforts to protect places like this and the Great Lakes themselves. I like to think that "Pictured Rocks 101" was part of his motivation for becoming an eloquent defender of the lakes in prose. He says it was. In fact, decades later he said, "Waking up to the sound of Superior waves slamming the base of the bluff and looking out over seemingly infinite blue waters to the north, I knew this was my life's work."

On the same Lakeshore trail, I introduced Carl Zichella, a new Sierra Club colleague from California, to Lake Superior on a crisp autumn day. Hiking along the escarpment, we stopped to rest on top of the cliffs at one of the viewpoints, and a huge flock of Canada geese, pushing hard across the breadth of Lake Superior, flew right over our heads only a few feet above us.

The sound of wing beats and goose call suddenly surrounded us, and Carl and I stood transfixed. We could feel the down draft of their wings and the great noise and activity, and then just as suddenly they were gone. Their raucous voices faded in the sky, leaving us alone on the cliff with the wind, the lake, and the stillness. Wild places are filled with moments of grace and wonder like that, and they are gifts made indelible in memory to last a lifetime.

After that welcome, Carl and I talked in the campground, which we had to ourselves. The lake was quiet, and we could hear the gentle rush of the Mosquito River running over its shallow steps into the lake. Carl said that places like this reminded him of the lines from the 23rd Psalm and that this recollection surprised him, because he didn't think of himself as being particularly religious in the traditional sense. Spiritual beauty speaks to us in many ways, and it isn't all that surprising that a psalm so widely known in Western culture would percolate up to form such an association in a place like this.

That evening, as the light left the sky on that trip with Carl, I sat on the little bridge over the river, and the psalm played in my head, forming my own prayer of gratitude. Thank you for leading me beside the running streams and the green forests. We are truly overflowing with beauty in places like this. In this beauty, I am in the presence of goodness and mercy for the weary soul, and in a place like this, I know that I am dwelling in the gifts of a benevolent creator who loves beauty, life, and the fullness of his or her creation. Thank you, oh Creator, and big bang, glaciers, time, gravity, and water for the gift that is Pictured Rocks. Thank you, too, for the hardworking activists, politicians, taxpayers, and anyone else who made it possible for me to come here to restore my soul from time to time.

Now, all these years later, I was feeling that same sense of gratitude to be back at the lakeshore again. But while I indulged in memories and reflections, the low angle of the light on the water told me the afternoon was wearing on, and I knew it was time to leave. I glanced across the river to see my companions readying for the return hike. I tucked my memories away in a safe place and pledged to come back to this piece of earth that knows me so well, sooner next time.

Reluctantly, we gathered up a waterlogged child, shoved more feet into grudging socks, and headed back to the car. The leaving is always a harder walk than the coming, but I was satisfied when my tired young hiker said "it was worth it" when we talked of the pleasures of Mosquito Beach on a sleepy ride back to our rented cabin.

Welcomed back by the lakeshore after too long an absence, I was eager for more time there. The next year, we returned with old friends and new. Colin's best friend, Eben, and his mom, Lynn, joined us for a few days, and Colin was thrilled to have a buddy along. We were eager to show them places that would become memories.

On a gray and drizzling morning, we packed our rain gear and squeezed everyone into the car. Our destination was a short hike from a different trailhead back into the Mosquito River Basin. By the time we reached the trailhead for the Mosquito and Chapel River Basins, the weather had shifted to a light mist and things were looking up.

The boys were skeptical about the adult-planned death march through the woods in the rain, but they were outnumbered, and we loaded up the daypacks and headed down the trail. Our path loosely followed the course of the Mosquito River down to the lake. Refreshed by rain, the forest was glowing with green mossy banks, and we traveled under the canopy of grand old trees. The walk was downhill and easygoing, and the boys armed themselves with good hiking sticks discovered at the trailhead (which also served as impromptu fencing foils as needed). They kept up without being cajoled (another good sign).

After half an hour or so, we heard the rush of the river on our left and were soon at the edge of the campground area. Moments later, we climbed up a small rise to the bluff sheltered by white pines and offered Eben and Lynn the view of Lake Superior. Even with mixed clouds, there were splashes of blue in the sky, which produced intense reflected color in the lake. From the expressions on their faces, it was clear this destination was a hit.

We eased down the log "stairs" to the sandstone ledge and spread out for a picnic. We nibbled on our simple lunch fare and took in the expanse of the lake and the pleasure of this place. Eben and Colin set about high grading the gorp to select out the M&Ms. This is a breach of our family camping ethics on a longer trip where those sweet treats become precious, but it hardly seemed worth an intervention for one afternoon. In the unsettled sky, blue patches were breaking through the billowed rain clouds. The sunlight was gaining, brightening the lake with deeper blues that showed off the contrast of lake, shore, and forest.

After lunch, we crossed the bridge and scrambled down to the familiar riverside beach. The boys were drawn into the water as if it had a magnetic pull, and they found a warm spot where a shallow eddy near the river mouth had not mixed with the chilly river or big lake waters. They played in the waves, and their pockets filled with sand. Sitting on the bottom, facing the waves, they got a full-frontal whack with each break, which sent them tumbling and screaming until they braced for the next set. It is good to be ten. Bill wandered off with his binoculars to see what the forest might reveal, and Lynn and I kept an eye and ear on the boys while inspecting the rocks, large and small, on the beach.

Lynn was taken with a glossy dark rock by the river, and she drew my attention to the way it felt under our palms. It was smooth, with an elegant patina. It reminded me of the soft black of ebony keys on an old piano. She guessed that it was basalt—ancient lava rolled and polished to a compact bench for our lifeguard duty. Her powers of observation revealed a new intrigue in a favorite place, and I watched and learned as she pointed out new discoveries on the beach.

The beaches of Lake Superior along Pictured Rocks are born of sandstone, and between Pictured Rocks and where the substrate changes to granite near Marquette, the shores are often strewn with candy-striped rust-red and creamy white ovals of Jacobsville sandstone. I found a near-perfect "yin and yang" stone with a dot of red on the white swoop, and a dot of white on the opposite red swoop. I longed to bring it home, but I decided that this is where it belongs, and perhaps it will fascinate someone else on another beach walk. I let it sink back into the kaleidoscope of multicolored stones and pebbles slipping in and out of focus under the lapping waves.

Lynn's pocket field guide revealed that our handfuls of pretty stones included gneiss, quartz, gabbro, more basalt, and a piece of jasper or two along with all the sandstone. Although the lakeshore is sculpted from ancient layers of sandstone, the comings and goings of the glaciers scraped and scoured the entire northern half of the continent, scattering boulders and pebbles from multiple layers of the earth and countless locales across the landscape on their retreat. As a result, our hands held samples drawn from several billion years of Earth's history. Like gazing at the stars, such knowledge can make one feel like a tiny presence in the universe.

Bill came back to take over as the watchful eye on the boys, and I wandered off for a bit by myself to indulge again in a quiet moment in the old campsite area on the bluff. I stood under a big white pine, taking in the view, trying to soak it up and store as much as I could, not knowing when we'd be back again. I grew wistful. I knew that my place of sparkling waters and green forests was someone else's view of a lost opportunity for development. And in this context, my relationship with places like this brings me joy but also sadness for the great divides in our cultural values that make the protection of the sacred and beautiful the stuff of contentious and hurtful disputes.

I returned to Pictured Rocks because in its wholeness and beauty it is a special place that speaks to my human spirit, as I think it does to others. That makes our enjoyment far beyond play, although I would argue that genuine play is something that a lot of our society is starved for. Our enjoyment comes from feeling part of this place, feeling the rhythm of the seasons and even deep time written in the history of the sandstone cliffs. We stand witness on these shores to the earth's biggest expanse of fresh water, tasting the freshness of the air, watching clear water lap over our toes. These are riches that never show up on a bank statement or a gross national product index.

The Anishinaabeg lived in Lake Superior country, and honored the spirits along these shores long before Father Marquette himself viewed them with awe and even longer before my ancestors settled in Marquette, Michigan. Surely the sacred beauty of this place is evident—palpable to anyone who comes here with an open mind and heart.

At Pictured Rocks, all around us we can see the forces of nature, God, or the spirits of the Manitou and the four winds at work. We could reduce it to mere forces of geology—a juxtaposition of rock and water with a little forest and wildlife thrown in—wondrous enough in a sense. But humans are capable of seeing more and knowing more than fact and measurement. We are heart as well as head, poets as well as scientists.

The ineffable is as important as that which we claim can be "known." As Shakespeare's Hamlet points out to his loyal friend, "There are more things in heaven and earth, Horatio, than are dreamt of in your philosophy." And for those of us who are curious about what lies beyond our philosophies and open to the "more," a deeper connection awaits.

Long ago, on one of the congressional tours of potential Wilderness areas, I stood next to Congressman Dale Kildee on another beautiful Great Lakes shore. He looked across the sweep of the sparkling bay and the cedar-lined shore, and said, "When I think of Michigan wilderness, this is what it looks like." Then, as he was wrapping up his comments to the group, talking about how these areas were largely unchanged by human hands since the time of the glaciers, he thoughtfully said, "When you think of it, wilderness is the mind of God expressed." Amen to that.

That expression of the mind of God, or, if you prefer, the evident progression of evolution and geology, is alive and ever-changing and has much to teach us. In places of extraordinary beauty, when we slow down and open ourselves to experience it, it washes over us like a holy balm. In lake country, as the poet Yeats put it, "I shall have some peace there, for peace comes dropping slow."

Every inch of Pictured Rocks National Lakeshore is precious in this world where so much is paved, developed, and manicured. This wildness was hard fought and suffered for. The place is fraught with compromise. People raged against leaving it alone. Some clamored for roads, resorts, and economic development—the whole package.

I understood the yearning and need for new income, and a better tax base for the schools—I still understand. Some people hated me and my ilk for wanting to save this remnant of the wild shore for its wild self. It hurt then, and maybe it still does. But my heart ached to save what we could of the wild spirit and beauty of this place, and I followed the truth in my heart.

The spirit of Anishinaabewi-gichicami—whispers in the pines and in the deep, dark waters of the greatest lake. I know that so many others have heard that voice as I have, and their spirits, like mine, are woven into the sacred fabric that is this place. I am grateful down to my bone marrow that I have known this place, and that untold others will know and love it too.

PART FIVE

Reaping What We Have Sown

Therefore they shall eat the fruit of their way,
and have their fill of their own devices.

—Proverbs 1:31 ESV

CHAPTER SIXTEEN

Still Toxic after All These Years

In the summer of 2005, I did something radical. I looked over the menu at Au Train's Brownstone Inn and ordered Lake Superior lake trout, deep fried. When the meal arrived, I looked at this crisp golden fillet—the first lake trout I had knowingly eaten in more than twenty years. I raised my fork, hesitated for a moment, and then plunged in. It was *so* good.

Lake trout are the native top predators of Lakes Michigan, Superior, and Huron. They are salmonids, an ancient fish family, which includes salmon, trout, and grayling—some of nature's most beautiful and delectable fish. They all share an affinity for clear, cold, oxygen-rich water. The lake trout's Latin name is *Salvelinus namaycush*. The genus *Salvelinus* means "little salmon," and it includes charismatic cousins such as the arctic char and brook trout. The "little" part might be misleading, as these long-lived fish can grow to impressive weights well over thirty pounds.

Their species name is *namaycush*, a warm and fuzzy word that in my English-speaking mind connotes something cute and cuddly like a stuffed animal, which is not what a self-respecting lake trout resembles at all. The Ojibwe word is *namegos*, and there are close cognates in other Algonquin languages. Several sources suggest it means "dweller of the deep," which certainly describes its habit of spending most of its life in very deep, cold waters. Other sources claim that namaycush means "tyrant of the lakes," which makes sense for a top predator known to eat at all levels in the food chain. Both descriptions fit. Namaycush is also called a laker, salmon trout, or gray trout. The laker is the lean member of the lake trout family (still carrying an impressive 20 percent body fat around in those chilly waters). The husky version that lives deep in Lake Superior is called a siscowet (or siskowet, or other spellings), whose body is about 60 percent lipid, or fat. The name is said to translate to "cooks itself" in Anishinaabemowin (Ojibwe language), and you can see why.

Given all the tribulations of the lake trout over the last fifty years, it was something of a miracle that a commercially caught beauty was on the menu in the first place. I was delighted just to have the choice of eating one. Doing so was a radical act for me because I know what accumulates in the flesh of a lake trout, and I had heeded warnings not to eat them during my childbearing years since I first learned about them in the 1980s. By that summer in the restaurant, I was reasonably confident that my childbearing years were behind me. Moreover, this was a Lake Superior lake trout, and thus likely less contaminated than from any other Great Lake source, given the high quality of the big lake's waters. I also presumed that it was of wild stock and from as robust a lineage of lake trout as can be found anywhere in the Great Lakes.

I considered that what remains of the commercial fishery is highly regulated as to what it can catch. I expected that the kitchen had removed most of the fatty skin and dark brown flesh when they filleted this trout and prepared it for cooking, and that if they hadn't, I would be able to easily set it aside. Finally, I considered that PCBs—the most likely contaminant in the fish—are mid-level carcinogens (the potential reproductive and developmental effects are much more serious). Thus, they were unlikely to cause trouble for me—at least not for decades, and this was just one little dose.

Twenty-five years had passed since PCB manufacture was banned, cleanups were underway in some of the most contaminated sites, and there were signs that ambient levels in the lakes were declining. While I was indulging, I opted for deep fried without recalling that hot temperatures will melt any untrimmed fatty tissues and leave whatever was stored in them in the deep fryer to contaminate all the other food cooked in the same oil. Bon appétit! Is it just me, or is this a lousy way to have to study a menu? I resent having to calculate a risk assessment between the "I'll be your server" greeting and the "Are you ready to order?" moment. "Yes," I'd love to reply, "I'll have a fresh, healthy, native lake trout fillet, but could you ask civilization to hold the toxics? I'd be so obliged." My attitude is showing.

The previous summer, our son had his first opportunity to go fishing in a northern lake. His uncle Rick was the patient teacher as they went through the basics. They stood in the shallows of a beautiful little U.P. lake and started casting. When Colin caught his first bluegill, we had one proud and happy kid. Rick showed him how to take it off the hook and gently release it into the water. Colin just loved the whole experience (except the worm on the hook part, and I can sympathize). On his second day of fishing, he caught a fish large enough to be "a keeper." I was raised in the "you caught it, you clean it" approach to fishing and had no desire to go through the effort it would take to scale and gut this fish, but I was thinking about it as a way of honoring my son's enthusiasm. Then I remembered that he was a developing child, and this was an inland lake that was probably mercury contaminated. While I didn't think bluegills were on the advisory list, I didn't have the complex booklet with all the guidelines and warnings handy, and I just didn't want to regret this decision later.

Our family doesn't eat a lot of meat, but we're not vegetarians. I figure if you are going to be a meat-eater, you should understand what it means to take a life for food and respect its gift to you if you do, and this is part of what fishing for "keepers" is about. So, I found myself in this moment of my son's pride and triumph explaining that it might not be safe for him to eat this fish, and I watched his face fall. We let it go, along with his first opportunity to take pride in the act of providing his own sustenance. When a kid can't go fishing with his uncle and taste the fruits of his labors, something is very wrong in Great Lakes country.

The contaminated food web is such a stinking crime. It is a crime against the fish, the loons, and the eagles; a crime against the people who made their livelihoods in commercial fishing; a crime against the people who depend on these fish for sustenance; a crime against humanity for placing so much valuable protein off limits in a hungry world; and a crime that a daughter of the lakes has to wonder what she's doing to her developing fetus or infant if she's had a few too many fish.

Part of my mama-bear caution on fish contaminants was a result of a summer's day in 1989 that I spent with Jim Ludwig and his research associate, Hiroko Kurita, while they collected data on a Lake Michigan island. They were studying failed hatches and birth defects among birds that live in colonies on Great Lakes islands. Their work contributed to mounting evidence that contaminants in the Great Lakes were linked to very troubling trends in fish-eating birds. We spent the day in a cormorant colony as they examined the evidence and took notes.

Many fishermen hate cormorants because they are convinced that these efficient diving birds are competing with them for Great Lakes sport fish, and they'd just as soon see them gone. But that day we weren't there to judge the cormorant's role in the Great Lakes fishery. We were just identifying and assessing eggs that hadn't produced successful embryos and documenting the chicks with physical abnormalities.

There's nothing glamorous about this research, walking along guano-covered rocky shores, examining smelly, often rotten eggs, documenting a variety of things that just aren't normal in failed embryos. I remember Jim sitting on a rock, methodically stating what he was finding while Hiroko recorded his comments in a notebook. "Edema," he would say, "failure of the abdominal wall to close," and other descriptions of things that you don't want to hear associated with wildlife in the middle of Lake Michigan. By the end of the day, we had found three little chicks with bills that wouldn't close properly.

The upper and lower portions (mandibles) twisted or crisscrossed around each other, which meant that these birds would never be able to dive, fish, and fend for themselves. It was common enough in some areas that it was being called "cross-billed syndrome." Researchers had also found this abnormality in terns. In humans, the parallel abnormality would be expressed as cleft palate. Jim said that cross-billed chicks who survive their early weeks will typically fail when they attempt to fish for themselves. The instinct to dive for food is strong, and they try it, but because they can't close their bills, they drown.

Given that the cross-billed chicks were doomed in their natural habitat, he opted to bring them back for observation and eventual analysis. (Cosmos—the cormorant Jim had brought to Washington, D.C., in 1989—had lived for fourteen months under Jim's care and observation.) Back at the boat landing, I helped Jim unload the boat and he handed me one of the little fuzzy chicks. I held this trembling cormorant chick with its horribly curled and twisted bill, knowing it would never survive, and looked into its little dark eyes. How can we account for what the chemicals of our world have done to it and the other innocent lives in the lakes? It still haunts me. Failed eggshells in eagles in one decade; crossed bills in cormorants in another. Both are avoidable crimes against nature. You wonder when we'll learn to consider the consequences of what we use, dump, and flush, and realize the lakes are our lives, not a dumping ground.

The actual dumping grounds had their own nightmares to contend with. In the spring of 1987, I attended a meeting in the Buffalo/Niagara region that included a bus tour of nearby Love Canal. Local community leaders organized the tour, including activist nuns in the Ecumenical Task Force of the Niagara Frontier. Our tour hosts included some whose empty homes in the contaminated zone were now behind a chain-link fence. That fence surrounded a neighborhood built on soils dense with PCBs, dioxin compounds, and pesticides dumped by the Hooker Chemicals and Plastics Corporation in the 1940s and 1950s.

Pat Brown was the host on our bus, and she asked the driver to stop so we could look through the fence at the red tulips she had planted near her front door—a place she would likely never be able to visit again. This, too, was haunting. I drafted the following poem, "Love Canal, for Melanie, after the Tour," afterward for my Washington, D.C.–based colleague, Melanie Griffin, who had joined me at the conference to learn some of the ropes of the Great Lakes and toxics issues. I also wanted to honor Pat Brown, Sister Margeen Hoffman, Lois Gibbs, and so many courageous others who were fighting for this community and the larger environment.

I saw the ghosts at Love Canal
And heard the echoes of children.
Playing in empty streets,
Empty yards,
Boarded houses.

Did the guilty weep great heaving sobs into their pillows
For ruined lives,
Shortened lives,
Lives that stopped before they even started?

I saw Pat's tulips blooming in the poisoned soil
And I saw Pat's eyes
And I will not forget.

And so, to leave despair for courage,
Anger for conviction,
Anguish for hope,
I will bear witness to the tragedy
And the seeds of triumph,
And I will act.

In Dante's version of hell in his famous *Inferno* poem, one of the lowest and most terrible places is reserved for perpetrators of crimes against nature. He had a different set of "offenses" in mind in his time, but surely, contaminating an entire ecosystem's food web and the soil beneath neighborhoods where humans live and breathe is a crime against nature that deserves searching and testing of the soul. Dante's supreme being doesn't cut much slack for crimes committed by inadvertent or unintentional action. Will ours?

Nobody set out to ruin the fishery. I'll give the chemists who first synthesized PCBs in 1880 a small break, along with the folks who rapidly snatched them up because of their useful industrial and electrical applications. But they didn't look at what else these compounds could do when they escaped the electrical transformers, manufacturing processes, or brake linings. No one apparently asked, "What could go wrong?" The same goes for many of the rest of the chemicals on the IJC's critical pollutants list that first caught my attention in the 1980s. More than a century after the advent of PCBs, with eighty thousand chemicals already in commerce, the United States still unleashes an average of two thousand new chemicals into the environment every year without asking that question.[1]

Human health studies are relatively scarce in this field. This research is expensive, requires careful ethical consideration (you can't just dose someone with a toxic chemical and observe), and it can take decades to look at human growth and development from fertilization to adulthood and beyond. So, researchers look at human umbilical cord serum, blood, and breast milk, or interview and observe people who may have been exposed to a particular substance. Researchers also use animal studies, given that humans are animals, and biochemical reactions in our living cells and hormones aren't that different from other life in the animal kingdom in our food chain.

In 1984, the *Journal of Pediatrics* published a study led by Joe and Sandra Jacobson that looked at children of mothers in western Michigan who regularly ate contaminated Lake Michigan fish.[2] The newborns weighed less and had smaller head circumferences than unexposed babies, which the authors described as "worrisome." This led to other work in the field, and the Jacobsons and colleagues followed these children as they grew up, noting subtle but measurable cognitive effects, such as "less efficient visual discrimination processing and more errors in short-term memory scanning" among four-year-olds.[3] A different team looking at nine-year-olds concluded in 2008: "These results, in combination with . . . a similar study in the Great Lakes conducted ten years earlier, indicate that prenatal PCB exposure in the Great Lakes region is associated with lower IQ in children."[4] There are also indications that in older adults, PCBs affect memory. In 1996, Theo Colborn, Dianne Dumanoski, and John Peterson Myers published *Our Stolen Future: Are We Threatening Our Fertility, Intelligence, and Survival? A Scientific Detective Story*. It summarizes trends in research and explains how PCBs, dioxin, and DDE influence hormones and the endocrine system. This affects fertility, development, immune health, and likely behavioral health. It is a sobering read.

Today, scientists classify pollutants that affect these systems as endocrine disrupters—chemicals that disrupt the normal function of the endocrine (hormone) system. One of the times in your life cycle that you really want your hormones to be sending the right messages to the right destinations is during reproduction and development. Thus, one of the big concerns about these chemicals is their potential impacts on moms who will bear children and the developing children themselves.

With lipophilic toxic chemicals, that fat-loving characteristic comes into play in troubling ways in the mother and child relationship. Unfortunately, not all fat is used for long-term storage, and the lipophilic compounds also end up in breast milk, or more subtle places such as the myelin sheath that wraps around our nerves, or ovarian fluids and other destinations where you'd rather not have DDE or PCBs as a close bedfellow. Moreover, most of these chemicals can and do cross the placenta. A woman's most efficient way of reducing her body burden of these chemicals is to have a child and nurse it—a terrible fact. The rich composition of breast milk makes good use of fats from a mom's body. It also brings along an unwanted "welcome to the world" bonus of what mom has been storing up in her tissues all her life. While there's some good news that PCB levels in the predator fish have been slowly declining (estimates are about 3 to 7 percent a year, depending on the lake and species), at this rate we have decades to go before we can declare "clean and safe."[5]

Let's not forget methylmercury, which also crosses the placenta and can be present in breast milk. It is also a reproductive and developmental toxicant, and can affect the immune system, nerve function, and brain. The eighteenth- and nineteenth-century "mad hatters" were so named because these hat makers were poisoned by the mercury nitrate in the felting process and their symptoms imitated insanity. We also learned from the tragedy of the industrial contamination of Japan's Minamata Harbor in the 1950s what methylmercury could do to humans when there are high concentrations in the marine food web. The suite of symptoms, from brain lesions to diminished sensory capacity to tremors (and beyond), became known as Minamata disease, affecting more than two thousand people over more than thirty years.[6]

In lower concentrations, like what most of us experience with eating fish, we won't see the effects of the hatters or the people of Minamata, but you just don't want to mess around with this stuff if you have a choice. However, when it comes to fish in today's world, you don't entirely have a choice. The number of lakes under fish consumption advisories for mercury far exceeds those for any other contaminant on the list. Advisory guidelines for safe eating apply to fish from all inland waters in my home state of Wisconsin, "because fish from most waters contain mercury."[7] For men and women over fifty, small panfish and inland trout are on the unrestricted list, except for waters with high mercury levels. How's the rain in your state or favorite ocean?

The human body can metabolize and excrete very small amounts of mercury. My mercury-savvy colleagues explain that its "half-life" in the human body averages between thirty and ninety days. This means that half of the mercury you ate on day one is probably out of your system between one month and three months later. The rest? Half of that will be gone on the same timeline, and on and on as you slowly work this stuff out of your body. Averages mean, of course, that it can take longer for some people. As a result, my friend Eric Uram, a passionate angler who knows his chemistry, suggests that a woman should be carefully watching her fish intake for about a year before a planned pregnancy if she likes to eat fish—especially if her favorite fish include those species that pack a mercury wallop. Because we can successfully get rid of *some* of this stuff in less than a geological epoch, I continue to selectively eat tuna. I love the taste of tuna (especially a cheddar tuna-melt with a real summer tomato) and assume I benefit from all the other good things in its oil-rich flesh. But I don't eat it very often. Whether it is ocean fish like tuna or swordfish, or freshwater fish like lake trout or walleye, any tasty predator fish at or near the top of its food web will have mercury in its flesh.

If you don't eat fish, don't assume this is not your problem. Some farm animal feeds include protein derived from ocean fish (which is what most fish farms are feeding carnivore fish). If you do like fish, eat strategically. Eat occasionally, or lower on the food web, and opt for fish with less fats and oils from nonindustrial, remote areas. However, the jury is still out on the best strategy to avoid PFAS, as they are showing up in smaller fish at very concerning levels. So, one approach does not fit all chemicals or fish. (This is where the vegetarians come out ahead in the bioaccumulation game.) If you crave salmon, wild Alaskan salmon is your best bet.

Most experts agree that the positives in breast milk outweigh the negatives of any contaminants, and I didn't hesitate to nurse my own child. But still, dosing your wee one in the rocking chair is a bitter thing to lay on the consciences of the mothers of this world. People who advocate market forces for cleaning up the environment talk about requiring the polluters to compensate the affected parties, on the assumption that the high cost of doing so would provide a deterrent to future chemical releases. I sometimes joke that I am still waiting for my compensation check for serving as an unwilling storage facility for by-products of the chemical age, and now they owe another generation compensation too. In truth, I would rather have clean fish for all future mothers in perpetuity. How much is that worth?

I remember sitting at a conference dinner long ago with a male colleague who wanted to know whether the Great Lakes fish on the table was safe to eat. I said, "Are you and your wife done having children?" and he replied that his kids were grown and in college. "Go for it," I said, and handed him my plate as well, and he finished it off. Years later, my lake trout fillet at the restaurant probably carried some trace combination of nearly all the banned chemicals, some mercury, and a potentially significant kick of toxaphene when I popped that first delicious forkful in my mouth, but I'm relatively confident it was a far lower dose than it would have been twenty-five or thirty years ago. Some comfort, isn't it?

Back when I was lobbying for the CAA, scientists told me not to expect to eat Great Lakes fish like lake trout with abandon again in my lifetime, and the same is likely to apply to the next generation. This is further evidence of how persistent these chemicals are in the living environment and how inadequate our response has been. Let me state for the record that I am not a live-in-a-glass-bubble, everything-causes-cancer, and we're-all-going-to-die-tomorrow type. I don't live my life in constant fear of what's going to get me next, and I don't recommend anyone else embrace that approach either. But I am deeply affronted that this stuff is in my lakes, my food, and my body without my permission. And it isn't just about me, or other moms, for that matter. For some people, fishing is about heritage, sustenance, or survival. And, for the non-human species in the aquatic ecosystem, tofu isn't an option.

What troubles me the most is that so many have come to accept a contaminated food web as normal, as "just the way things are." In the summer of 2006, the U.S. EPA proudly announced that the Oswego River in New York was being "delisted" as one of the Great Lakes Areas of Concern. To be "delisted," a site must have restored its beneficial uses—fishing, swimming, and habitat. This is a huge achievement in environmental cleanup. I asked colleagues preparing for the announcement, "Has it really been restored?" Their pride was palpable: "Yes, this has been an incredible recovery—definitely worth celebrating—except for the fish being safe to eat. That one isn't really achievable right now." This nagged at me. The next day, I mentioned this to another colleague, who remarked, "If we held out for being able to safely eat all the fish, we'd never get these areas off the list!"

In fairness, the fish contamination is systemwide, so holding back recognition for locally successful cleanup feels stingy. Still, these polluted harbors have not only contaminated local fish and wildlife but are toxic vectors within the wider ecosystem and its food web, through mechanisms such as volatilization, currents that pull polluted waters and sediments into deeper waters, activity from benthic organisms, dredging, and through the wide-ranging predators distributing the contamination of their prey. Delisting shouldn't gloss over the status of fish contamination, lest this long-term challenge fade out of sight and out of mind. We should not claim victory when we have not yet won the war, and I worry about accepting perpetual contamination as the new baseline for an acceptable "normal." I'm also troubled when my cheerful government tells people that the benefits of eating contaminated ocean fish outweigh the risks, without so much as a passing mention of what we might do to make this unpleasant balancing of risks and benefits go away by cleaning up local and global sources of pollution.

Although the 1987 GLWQA included the critical pollutants list, as well as 272 known and 108 potentially hazardous pollutants, negotiators for the 2012 agreement scrapped all the lists. Their rationale was that the list was too big and unwieldy. The result was no list, no priorities, no pressure to do anything. Moreover, the new "Chemicals of Mutual Concern" identification process is no longer led by the SAB and includes industry representatives on the committee. As of 2022, the list only includes eight chemicals: two fire retardants—hexabromocyclododecane (HBCD) and polybrominated diphenyl ethers (PBDEs)—good old mercury and PCBs, and four other compounds—long-chain perfluorinated carboxylic acids (LC-PFCAs), perfluorooctanoic acid (PFOA), perfluorooctane sulfonate (PFOS), and short-chain chlorinated paraffins.[8] Can you spot the chlorine, fluorine, and bromine in the compounds?

The newer chemicals all bioaccumulate, and levels of PBDEs "in fish like walleye and lake trout, along with sediment and gull eggs, exceeded safe guidelines and show no clear evidence of declining at this point; the same is true for the perfluorinated compounds."[9] Keep checking those fish advisories before you eat fish. And keep asking your elected officials and regulatory agencies why we have failed to adopt a precautionary approach to industrial chemicals.

A 2017 IJC web post noted a draft report on PCBs that stated the following: "despite initial steep downward trends in the 1980s and early 90s, followed by a much slower decline in PCB concentrations in fish tissues and sediments, concentrations still routinely exceed relevant guidelines, drive fish consumption advisories in all five lakes, and therefore pose a threat to the Great Lakes environment and to human health in the basin."[10] It also stated that "mercury is still at threatening levels to the environment and human health in the Great Lakes," referencing a 2015 IJC report that further noted:

> The Great Lakes region lacks an adequately and sustainably funded, geographically diverse and consistently maintained mercury atmospheric deposition monitoring network. Meanwhile, mercury levels in some fish and wildlife species in some locations of the Great Lakes Basin region are increasing.[11]

So, we don't even have sufficient research funding to develop the capacity to really track this stuff, and in some places, levels are increasing. While we're finally making progress on U.S. emissions, global sources continue to bring us toxic gifts on the wind. Toxic contaminants will remain a threat for decades, especially if we fail to reorient our regulatory system to prevent harm and fail to embrace precautionary approaches. Meanwhile, around the world, as industry continues to introduce new variations of persistent chemicals, they will continue to enter the environment and join the list with the near-forever old ones.

CHAPTER SEVENTEEN

Whither the Fish

As long as there are still viable whitefish populations for commercial fishing, I will eat them with much less concern than lake trout or tuna. I love their delicate taste and had forgone them for much of the last twenty years, figuring I had been megadosed as a child and teenager because my mother loved whitefish and lake trout when they were still available, and served them often. Whitefish have a flaky, less oil-rich flesh and they eat lower on the food chain than lake trout or salmon, thus accumulating fewer contaminants.

Whitefish love to eat *Diporeia*, a tiny freshwater shrimp-like critter that you and I have probably never seen. Invasive species have been aggressively eating and crowding out the *Diporeia* for years. Invasive filter feeders such as zebra mussels and quagga mussels suck in and digest just about every tiny thing in the water and don't care whether they are snatching some other species' food supply in the process. Then, there's the spiny water flea. If you are an unsuspecting whitefish nibbling at little stuff in the water and you go for a spiny water flea, the spines will jab you in the mouth and you'll spit the nasty thing out—but you'll still be hungry.

In addition to the toxic mess, we must deal with the invasive species nightmare as well, or one invader after another will keep tearing apart what is left of the food web and natural diversity in the Great Lakes. One of the ugly stepsisters of global trade is the unwanted invaders that ride into distant ports on the ballast water, packing crates, food shipments, or other vectors, from fish farms to the home aquarium trade. How many more emerald ash borers, zebra mussels, purple loosestrifes, and exotic viruses do we need in the Great Lakes before we realize that one of the costs of unrestricted trade and the open gateway of the St. Lawrence Seaway is havoc on local ecosystems? When ecosystems suffer, we pay.

In the Great Lakes, we have already spent billions of dollars trying to cope with the disruption caused by invasive species. We've made some progress, but we're still living with the legacy of the open-door policies. Ocean-going ships that come into the lakes (also known as "salties") now face strict ballast water management rules. Inbound salties typically carry ballast water in their holds for balance and stability—water pumped onboard from various ports around the world. Captains historically discharged this ballast water into local harbor waters to rebalance a ship's weight after loading on new cargo. Ballast water was the primary vector into the lakes for invaders such as the zebra and quagga mussels. Today, inbound salties must fully exchange their ballast water at sea before heading up the St. Lawrence Seaway, or treat it with chemicals that kill the organisms in their holding tank water, or keep it and residuals in their tanks onboard while in the lakes. It isn't a foolproof system, but it is making a difference in dramatically reducing new international marine hitchhikers.

Unfortunately, we can't go back in time. I came across this jaw dropping quote in a 2021 paper on how invasive mussels are influencing phosphorus in the Great Lakes: "In terms of biomass, quagga mussels are the dominant life form in the Great Lakes."[1] Sit with that a while—the dominant life form in the lakes. It took thousands of years after the Ice Age for the aquatic fish and wildlife in the Great

Lakes to evolve into the fishery we used to know prior to the era of invasive species. While there were early introductions of non-native species via the Erie and then Welland Canals, the invasion ramped up sixty years ago with the expansion of maritime shipping.[2] It took only twenty years for the quagga mussels to transform and dominate the system.

Sometimes the invaders find their way through their own means. Four species of carp (bighead, silver, black, and grass)—all native to Asia and often described together as "Asian carp"—were imported into to the southern United States in the 1970s to clean up algae in aquaculture ponds and wastewater treatment settling ponds. Whether it was from a flooded pond or a careless angler, bighead and silver carp found their way into the Mississippi River and headed upstream, including into river tributaries. Today, just a few decades down the road, bighead and silver carp are the dominant fish species in the Illinois River, which connects with the Des Plains River and, upstream, the Chicago Sanitary and Ship Canal. This provides a potential pathway to Lake Michigan via the Chicago River. If these fast-growing voracious plankton eaters find their way into the Great Lakes, these fish could wreak further havoc on the already disrupted Great Lakes food web.

Right now, electrical barriers in the Des Plains River at the Brandon Road Lock and Dam are the primary line of defense from a carp invasion. Many have called for fully separating the Great Lakes and Mississippi watersheds, which is how the glaciers left things a few thousand years ago. The engineered connection is only about 125 years old. Although it would be a massive reengineering endeavor, it is worth serious consideration.

The impact of invasive species on our already beleaguered fishery has been staggering. Researchers have documented substantial portions of Lake Michigan and Lake Huron that used to be great whitefish habitat that are now practically biological deserts, barren of the *Diporeia*. They have found whitefish with nothing but zebra mussels in their bellies, which the whitefish can't digest. The fish are just being starved out of big portions of their former range. You might just be tempted to tell those whitefish to buck it up and eat what's set before them. But we know that whitefish and evolution don't work that way. So, the beloved native whitefish gets whacked from a completely new angle, which some call biological pollution.

There are days when I despair for the native fish of the Great Lakes. Back in 2003, I attended a conference in Duluth where the focus was on aquatic biodiversity. In a session on Great Lakes fish, I asked two questions that I know are never popular. One was, From an ecological standpoint, would the lakes be better off without the Pacific salmon? The second was, Are the lake trout going to make it—do we see a day when they will be reproducing on their own and healthy again?

It was toward the end of the session, and these are the sorts of questions that make scientists who strive to live in the world of unbiased science extremely uncomfortable. Rather than take this on in open session, they asked me if I'd like to talk with them after the public session was over, so I did. Their answer to the first question was a qualified "we don't know." They explained that the aquatic biology of the Great Lakes is in such a mixed-up state that it is hard to know whether the salmon, which have filled so much of the predator role formerly played by the lake trout, are a plus or a minus in the open lakes. In the streams where the hatcheries release these fish, and in which more and more are beginning to breed on their own, they do compete for food and habitat with other native fish, which is problematic—especially if you are the native fish in question.

We didn't talk about the questionable wisdom of raising non-native fish that are aggressive, fast-growing predators, and as such build up some of the most impressive chemical body burdens of all the fish in the lakes. Nor did we discuss the fact that thousands of anglers take them home and eat them every summer or give them to their friends and neighbors who probably never even know that fish advisories exist. A salmon can reach the size of a ten-year-old lake trout in about three years, preying primarily on alewife (another oil-rich non-native fish). We're talkin' PCB vacuums here.

I decline to eat Great Lakes salmon, and you won't see them in your grocery stores because the commercial fisheries aren't licensed to catch fish raised for the sport fishery. Pay attention, though—salmon fish farming (which has its own set of ecological issues) is emerging in the Great Lakes region.

The second question was harder, and this is the one they probably didn't want to answer in a public session. We talked about the multiple efforts to restore the lake trout and everything that stood in the way. Before the invasives arrived, lake trout primarily ate lake herring, cisco, and other forage fish that the alewives have displaced in most of the lakes. Lake trout will eat alewives, but that's what the introduced salmon prefer, and devour with gusto.

At this juncture, the lake herring, and thus the lake trout, were holding on and doing fairly well in Lake Superior, and there were some hopeful signs in Lake Huron. Elsewhere, federal fish hatcheries stepped in to help keep the lake trout going, but as my friend Eric tells me, "This strained the genetics by mixing the lean lake trout with the deep-water, high-fat siscowet." As a result, some of today's catch for the dinner table no longer resembles the lean heritage of the original wild trout. Those who know their lake trout say the mixed progeny aren't as flavorful as the true lean trout. Improving nature's design for a wild fish is risky business.

It's hard, as the whitefish are discovering, to change your food source in a cold-water ecosystem when the food sources with which you evolved are no longer on the menu. This along with habitat and water quality are still significant factors. One scientist said there's been a ton of money spent on lake trout rehabilitation, with limited success. Another said the jury was still out. A third said there's been so much damage to the food web, he thought they'd never really return beyond Lake Superior. I can't tell you how sad I was after hearing these assessments.

I've talked to other scientists since who seemed more optimistic, and they point to Lake Superior and the relative success of lamprey controls. I also ran across a scary but hopeful 2006 article summarizing a paper that found a direct relationship between dioxin levels in Lake Ontario water and the complete failure of lake trout reproduction in that lake over a forty-year period from 1944 to 1984. That's the scary part. The good news is that since 1980, dioxin levels in Lake Ontario have dropped below the thirty-parts-per-trillion threshold. Aggressive pollution controls seem to being paying off. Maybe, I thought, just maybe, old *Salvelinus namaycush* has a chance. Another paper confirmed that hatchery-born lake trout were successfully breeding by themselves again in parts of Lake Ontario and Lake Huron.

Even the salmon—some of which are now reproducing in Great Lakes tributaries without the help of hatcheries, are finding that fluctuations in the alewife population make it hard to get by. In January 2006, the *Chicago Tribune* ran a story under the headline "SOS for Alewives" proposing that we should consider stocking alewives (in decline for a variety of reasons) to help the Pacific salmon. Raise one exotic to feed another? At taxpayer expense? It made my head swim. What have we done to the Great Lakes?

Not surprisingly, the alewife decline led to a decline in the salmon. Dan Egan, author of *The Death and Life of the Great Lakes*, posted a *Milwaukee Journal Sentinel* story in 2021 that got my attention. It explains that when the alewife crashed and chinook salmon declined in Lake Huron, native fish rebounded, including predators like lake trout, walleye, and smallmouth bass, as well as the fish they eat, such as chubs and emerald shiners. The story also talked about how the lake trout have begun to successfully prey on the invasive round gobies that live in rocky bottom areas. In the bizarre, highly altered Great Lakes food web, the invasive round gobies have the ability to crack the shells of zebra and quagga mussels and are one of the few things that can eat them. However, many of the predator fish—including lake trout, walleye, smallmouth bass, and even whitefish—can eat the round gobies, although it requires skill to snatch them out of their rocky hideaways.[3] Tenacious nature sometimes finds a way.

The return of functional natural processes is one of the best indicators of progress. A 2017 report noted that lake trout were reproducing naturally not only in Lakes Superior and Huron but also in Lakes Michigan and Ontario. Fantastic! (Lake Erie's were still all from hatcheries.)[4] In August 2021, *Field & Stream* reported that the New York Department of Environmental Conservation had discovered the presence of wild lake trout fry (babies) in Lake Erie for the first time since 1965.[5] Even more fantastic! So, there is hope. And a healthier, more naturally diverse aquatic ecosystem has a better chance of finding resilience in the face of long-standing and emerging threats.

Were it not for the expanding impacts of climate change, we could meet these signs of recovery with greater celebration. In the early 2020s, climate change has risen to the forefront of threats to Great Lakes fish and other aquatic life. As the lakes continue to warm, habitat will shrink for the cold-water fish—all the trout, the walleye, the whitefish, the exotic salmon, the cisco. Lake Superior is one of the fastest warming large lakes of the world.[6] A fish just can't get a break.

Reducing other stress on these amazing fish is essential while we attempt to tackle climate change. Every cleanup counts. We should celebrate every sign of recovery along the way. But I dream that someday we will not need to worry whether a wild fish in its original habitat can reproduce as nature intended. I also dream there will be rejoicing in the streets when the last fish advisory becomes an obsolete relic of the past. I hope a curator places the advisory booklets in a museum as a reminder of how bad it once was in the Great Lakes so that people will not forget what humans can do to an ecosystem.

If we are to atone for the twentieth century's crimes against nature, we must restore wholeness to the domain of *namaycush*. If we fail to try, Dante may have some suggestions as to what we deserve.

CHAPTER EIGHTEEN

When You Get Caught between the Moon and Detroit

In late 2005, I was on an afternoon flight from Columbus, Ohio, heading to Detroit's Metro Airport for the connection home to Madison, Wisconsin. It had been one of those glorious summer-like days of October. I suspected most of us in that full-day meeting were longing, just as I was, to be out walking down the sidewalks, kicking through the falling leaves, and enjoying warmth and sun before the dreariness of November. But we hunkered down over the task in front of us as good grown-ups do, and a rare golden day slipped by with us behind the glass, and me now in this sealed tube making the short dash to Detroit.

I've spent a lot of time going in and out of the Detroit airport, and the surrounding landscape and I seem like old friends. From a window seat, I can spot signposts on the ground that tell me I am homeward bound. On this trip, Lake Erie soon filled the view out the DC-9 windows, surrounded by the stretching farm fields of Ontario on the north and Toledo and its suburbs on the south. Erie was sparkling, all dressed up with no hint of her troubles, past or present—just blue water and bright sun. The Maumee River, draining western Ohio, was clearly visible, as was the Detroit River, Grosse Isle, Windsor, and, across the river, Detroit.

As we drew closer, the icons of downtown Detroit were easy to pick out. The "Ren Cen" towers are distinctive in their glassy cluster. Then there's the "golden tower of the Fisher building" from which WJR, "The Great Voice of the Great Lakes," has broadcast its strong radio signal since before I was born, indelibly branding that landmark with a tagline embedded in the brains of millions of Michiganders. The highways are distinctive—ribbons and little canyons of concrete squiggling in and around Detroit and spreading throughout the suburban counties.

From the air, the small scale of downtown compared to the vast spread of the suburbs is stark. Grid upon grid of houses reaches across the flat landscape in all directions, stopped only by the river and lakes to the east. But just beyond the concrete heart of the city and the gray highways were tree-lined streets in full fall splendor. The forest of sidewalk and backyard trees was glowing in the long rays of the waning sun, and from this vantage, the sprawling burbs of Detroit were simply gorgeous. I had to wrestle with this because suburban sprawl represents so much of what went wrong in Michigan with urban design, economic failures, and a thousand other things that shouldn't have happened to Detroit and other once-vital downtowns—including Flint, closer to home. But the beauty in my window invited me to set that all aside for a moment and delight in the tapestry: the rich garnet of ashes, amber gold maples and locusts, flaming red maples, and the russets that must be oaks. The remaining green in the landscape provided an emerald complement to all the other jewel colors. I was suddenly homesick for Pine Lake and autumn days where the wind played in the oaks and the colors across the lake matched what spread below me.

As if ground control knew, the plane shifted course, making a wide arc north of the city for a different landing approach. The lakes of southern Oakland County came into view, shimmering mirrors amid the multicolored landscape, and I knew that just a few dozen more miles north, Pine Lake was shining too. Seduced by beauty and memory, I was falling in love again with Michigan. But there was sadness behind the joy, because the show before me would be brief, and Michigan had trouble in her bones.

A little more than a week before, the Delphi Corporation had declared bankruptcy. Delphi used to be AC-Delco—a major parts division of General Motors. Over a decade-long process, GM consolidated AC Spark Plug, Delco, and various parts divisions into one unit, and eventually spun it off into the independent Delphi in 1999. The name appeared first in 1995 (snappier, admittedly, than the previous moniker, Automotive Components Group).

It isn't clear whether they intentionally borrowed the name from the home of the ancient Greek temple that housed the oracles who could offer glimpses of things to come, or whether it just sounded good to someone in a branding firm. Perhaps if they had consulted the oracles who were the original Delphi's claim to fame, they might have chosen a different course. But, as in many epic tragedies, our protagonists apparently neither asked nor heeded the predictions of those who could see the future. And so, a modern tragedy with a cast of fifty thousand workers was playing out as our pilot circled before landing.

I was taking Delphi's troubles personally because they affected people I love and a place that used to be home. My dad worked for AC Spark Plug for thirty-five years. The roof over our head, the food on the table, the cars in the garage, and my college education were possible due to my dad's hard work, the UAW, and General Motors. That unique collaboration was a slice of the postwar American Dream. My family was one of tens of thousands of families that bought into that dream and became the backbone of America's middle-class labor force. And now, for tens of thousands of workers and retirees, their struggling industry was scattering the crumbs of that dream to the wind.

The hand-wringing executives talked about the inability to compete with overseas wages, the huge burden of health insurance, the enormous pensions, and all the predictable bromides about global economies. The television pundits talked about the excessive benefits in the union contract, and they dropped bewildering hints that this labor force somehow overreached in its aspirations to a decent middle-class life. None of them were calling for universal health care that could lift the estimated $1,500 per car "health insurance burden" from the domestic employers. (The Japanese automakers don't carry this direct cost, as the nation has embraced universal health care since 1961.) No one was talking about a broken social compact between the labor force and American industry. No one was talking about trade agreements, such as the North American Free Trade Agreement or the global General Agreement on Tariffs and Trade. No one was talking about the lack of safety nets for the workers and the retirees and the fragile state of those that do exist. And no one was talking about the dinosaur decisions by the American auto industry. Was there no learning curve after the late 1970s?

While I am GM-born and bred, I could not embrace the loyalty to the corporation in the way my dad and so many of his fellow workers had for a lifetime. I saw the chilly way GM treated their workers when things got tough in the 1970s. Still, I remained loyal enough to drive and buy GM products in my twenties. But after two cars that had frightening safety and maintenance issues, my loyalty wore thin.

I was also butting heads with the auto industry on a law I'm very fond of—the CAA. I couldn't reconcile the union and the corporate leadership lobbying against their health and mine in the name of profits, and allegedly jobs. And so, with great trepidation, I sold my hated Pontiac T-1000 with the accelerator that randomly stuck at high speeds, and my husband and I bravely crossed the line in 1985. We bought a new Honda Civic. She had sleek lines, and we named her Emma after Mrs. Peel in the Avengers—smart, quick, nimble, and a saucy brunette. She purred like a kitten, got great gas mileage, was fun to drive, and never gave us an ounce of mechanical trouble. Even though Honda built our car in Ohio, it was outside the GM and UAW family. I felt guilty, but there was no going back. I was seduced by reliability, quality, and a corporation that seemed to understand that the environment was not the enemy but rather a marketing advantage. I was abandoning the corporation that had been my family's bread and butter. But the corporation didn't seem to care very much about my safety or health, or my planet, so I made my choice.

It is easy to drive a Honda in Madison, Wisconsin. The place is littered with them, and other small, efficient cars that are easy to park in a university town with narrow streets and more than forty thousand students. It was harder to drive it home to Michigan for a visit at Christmas. We'd heard the stories about people being threatened and harassed on Michigan freeways for driving "foreign" cars. But we mustered our courage, and my dad even let me park it in the drive. (I asked first.)

We totaled that sweet little car in 1991 on a slick intersection near Traverse City on the way to a biennial IJC meeting. On black ice, brakes aren't much help, and we skidded through a red light smack into the side of the car driven by a local postal delivery worker. He was shaken up but not seriously injured, for which we will remain grateful for all time. As for us, other than some seat belt bruises and the cuts where Bill's fingernails had clenched into his fist as he gripped the steering wheel, we walked away shaken but completely unharmed. The car performed exactly as it was supposed to in a front-end crash. The hood and engine looked like an accordion, but the passenger compartment was intact, and so were we. This locked in my new customer loyalty and offered me a response to the people who argued that big cars are safer on the road.

Suddenly in need of a new car, we didn't hesitate to get another Civic, although we waited a few weeks because air bags were coming out in the 1992 model. Somewhere a guy in Honda marketing was chalking us up as a "brand loyalty" success story, and somewhere a GM executive wasn't worried about writing off our "marginal" share of the market. We managed to wring more than a dozen years of service out of Margie, the 1992 Civic. She was a bit more troublesome than her predecessor (small cars go through a lot of mufflers with city driving) but still very dependable. However, around year eleven, rust, the cancer of northern cars, began to really take its toll, and her zip was showing signs of limited recovery. So, we started looking around again for a new car. There had been changes in the industry over that decade, and the GM family discount had its appeal. We thought we'd take a serious look at Saturns. Fate, the Big Three automakers, and the unions intervened.

Timing, folks, is everything.

Since I first became acquainted with the CAA, the world has pumped a lot of gaseous emissions and filth into the atmosphere. I doubt that anyone paid much attention to CO_2 emissions back when the original 1970 act was being crafted, but since then we've learned a lot about what this seemingly innocuous gas is doing to our atmosphere and thus our climate. So, just as I was getting brave about coming home to the Big Three, they had to get all huffy about CAFE standards.

CAFE standards sound like something a restaurant critic or the health department should care about, but the acronym stands for Corporate Average Fuel Economy. Congress first established CAFE standards in 1975 during the Arab oil-boycott years as a means of boosting fuel efficiency across the U.S. auto fleet. By requiring a minimum average fuel efficiency, the standards would conserve fuel and reduce auto emissions. The auto industry fought this tooth and nail but didn't prevail. The original goal was to double efficiency by 1985, which we achieved by getting the fleet average up to 27 miles per gallon. For the next twenty or so years, standards only increased to 27.5 miles per gallon, which hardly seems like a technological leap. In the early 2000s, Congress tried to raise the bar again. In addition to cutting down on tailpipe pollution, increasing efficiency across the fleet is also an effective way to reduce emissions of CO_2 and other greenhouse gases—good for the climate, good for our lungs.

Cars are a major part of the global warming problem. Period. Our cars emit tons of carbon dioxide, along with pollutants such as carbon monoxide, nitrous oxides, and particulate matter (a.k.a. soot). The kinds of cars we drive, the amount of stuff they emit, and the miles we run up on U.S. odometers every year are fixable problems. One practical step is to require across-the-board improvements in fuel economy. We've done this before and succeeded, but, just like the first round, the auto industry didn't want new requirements to be imposed in 2002 either. The domestic Big Three automakers were leading the charge to stop Congress from passing a new round of standards.

Becoming a parent heightened my sense of responsibility and conviction to work for environmental quality. As a parent, one sees the world through the eyes of what his or her child will inherit. A globally warmed planet was on my son's horizon, and I hated it, and he hated it.

People who do not live here believe that deep snow blankets Wisconsin all winter and that every winter weekend brings another Ice Bowl with the Green Bay Packers. These are legends from another time and largely nostalgia now. Nostalgia is insufficient for people who must explain to a seven-year-old why there is no snow again at Christmas, why we canceled the sledding party again, or why the Kites on Ice Festival is being held on land again. The climate has already changed significantly in the Great Lakes region. Those big snows we remember from our childhood *were* bigger, winters did last longer, and cold snaps were colder—although a new pattern now seems to be emerging: extreme snows in some years and just a trace in others.

During the winter of 2002—a few weeks after the Christmas when we had given our son his first sled, we had a lovely morning snow, but it was melting fast. By the time Colin was home from school, there wasn't enough snow to sled on after weeks of a winter drought. He flopped down on the rug, pumped his fists into the floor, and said, "I hate global warming, I hate global warming!"

So perhaps if he hadn't been moved to tears that winter, and if the Senate hadn't been considering a vote on auto fuel efficiency in March 2002, we'd be driving a GM car today. But, while we were weighing the big decision of what would be the next "ten-year car" in our family, my two senators and most of the other ones from the auto industry states got mushy. The UAW and the Big Three were lobbying aggressively, claiming that improving fuel efficiency across the fleet would cost too much per car and would threaten jobs. This was the same Greek chorus that we heard during the first and second acts of our tragedy. Apparently the chorus can't hear the oracle either.

I was mad. I'd expect this from the ordinary rank and file of the Congress. But two of the men I admired most in the U.S. Senate—Carl Levin of Michigan and Russ Feingold of Wisconsin, men with brains, integrity, and courage—bought the company line again. As if they hadn't lived through the 1970s along with the rest of us. As if making inefficient cars is good for business and good for you and me. As if the domestic auto industry is incapable of new thinking and bold leadership. And so, disempowered as a constituent in an auto state, I chose to cast my vote in another way.

I sent variations of the same letter to each of them, asking: "How can we mortgage our children's future and the earth they will inherit for the short-term gain of the auto industry, and how can a man of your stature and principles have led this effort? After a rare snow melted away quickly this winter, my seven-year-old broke down and cried because he thinks that the winters he loves are over—forever. He called it 'global *warning*.' He's a smart kid. No child should have to grieve about global climate change, nor should he have to inherit the legacy of our generation's callous disregard for the environment."

Then I explained, "I've already called our local Honda dealer and asked to be put on the waiting list to test-drive a new Civic hybrid. How it saddens me that the auto industry that brought great jobs to my dad's generation will once again punt the opportunity to be leaders in fuel efficiency and hand over consumers like me (and there are more and more of us) to the Japanese auto industry."

I didn't hear back from Senator Levin—members of Congress rarely respond to letters that aren't from their constituents. I get that. I got a generic reply from Senator Feingold's office that didn't really address my concerns. A senator from an auto industry state still marches to a certain drummer. I get that too, but I didn't have to like it.

So, we ordered a Honda hybrid and waited six months for delivery because we thought a hybrid was the right thing to do. We took 20,000 hard-earned dollars from a child of the Big Three and spent it on a product from a nonunion foreign corporation. And this twists in the heart. But the car, whose nickname was Silver Bullet, averaged forty-two miles per gallon in the city and lasted for fifteen years.

I value and honor the heart and soul of the American worker but also value the future of life on Earth. When the younger generations ask us how their elders could have seen the handwriting on the wall in terms of the threat of global warming and not acted, I want to be able to tell them that at least some of us tried, and tried *hard*, to do what we could to be part of the solution. Because my values didn't seem to matter to the domestic auto industry, because it couldn't seem to grasp that we can have jobs and innovation and a healthy environment, the domestic automakers forced me to choose between my child's future and my father's past—and that is a cruel choice.

It was dark when my plane lifted off for the connecting flight to Madison. The colors had faded to black, and there were only pinpoints of light below. A cold moon hung suspended in the night sky. My thoughts swam with the Delphi tragedy and what it would precipitate. I turned the big questions over and over in my head. How can we assert our rights to a new American Dream, where the labor force has decent jobs with good pay, high-quality health care, and a retirement safety net regardless of who employs us? How do we create a society where we do not pit employment against quality of life, and short-term need against long-term security? How can we fashion a world where global competition does not mean that other nations' labor forces must live on the margins amid environmental squalor while we consume more than our share of the world's resources?

We are a nation with great capacity and imagination. How can we wrestle opportunity out of the upheaval of a once-great industry? I rested my head on the chilly window and realized that I was pondering these questions on a bankrupt airline, pumping the sky with high carbon jet exhaust, and I closed my eyes and sighed.

Congress finally passed new CAFE standards in 2007 after gas prices soared above three dollars per gallon. The new standards, set for the 2011 model year, called for a minimum fleet average of thirty-five miles per gallon by 2020, which is less than our 2003 Honda already averaged. The new standards led to lighter materials, more efficient engines, and, for some of us, "donut tires" in the trunk instead of heavy spares. They also led to cleaner air.

In 2008, the Big Three asked for billions in low-cost loans from the public treasury to retool their product line. "Really?!" you want to ask. This caught you by surprise. Congress rallied to help the increasingly desperate domestic auto industry that summer, authorizing $25 billion for loans. Enter the home mortgage economic crisis, the hemorrhaging of Wall Street, and the dark days of that autumn. Just before Christmas, the Big Three went to Congress begging for more help at the scale of $15 to $25 billion. An angry Senate told them to file for bankruptcy, leaving the Bush White House to rescue what remained of the Big Three.

In 2017, we bought a silver Prius, and with her slick lines she looks like a fish, so we named her Namaycush. She can achieve about fifty-eight miles per gallon on the highway on a road trip, less in city driving. Personal vehicles are still a major cause of disruptive climate change—about one-fifth of all U.S. emissions. We pump about 19.5 pounds of CO_2 and other greenhouse gases (GHG) into the atmosphere for every gallon we burn (more if you are burning diesel), adding up to an annual average load from a single passenger vehicle of 4.6 metric tons.[1]

Climate change is not just another issue to worry about for the Great Lakes region. It is the overarching issue that makes everything else more difficult—warming waters that threaten viable habitat, exacerbate algal blooms, amplify lake-effect blizzards, and intensify thunderstorms that flush massive pulses of highly polluted runoff into the lakes. A changing climate also means extreme variability in lake levels that disrupts habitat and infrastructure, increased lake evaporation, and so much more. Every effort to delay action to reduce emissions is a further threat to us, our children, and the biosphere. Let us remember that we were warned by the scientific and environmental communities, the farmers and people whose villages were swallowed by the sea before we began to experience the massive forest fires, the countless "hundred-year" floods, the droughts, the heat waves, and the climbing temperatures. Still, our society chose to kick the can down the road. There is still time to act, but the window is rapidly closing. What will you tell *your* children and grandchildren?

PART SIX

Finding Our Way Forward

There are risks and costs to a program of action. But they are far less than the long-range risks and costs of comfortable inaction.

—John F. Kennedy, written message to 14th Annual Convention of Americans for Democratic Action, May 11, 1961

CHAPTER NINETEEN

Then and Now

More than three decades after our big achievements, the policy climate is divisive and stymied. This has constrained the region's capacity to advance safeguards for land, air, water, and people.

What happened between those "glory days" of political response to citizen activism, and today?

Riding high on the post-Vietnam progressive momentum, people organized, raised media attention, and lobbied, and the environmental movement built a federal policy framework. There were losses and compromises along the way, but there was also measurable progress. We connected scientific expertise with grassroots activism to shape and enact policies to clean up and safeguard the air and water, improve public health, and hold polluters accountable. We established ways to protect public lands from irreversible development or degradation, and we built greater understanding of living systems and our dependence upon them.

It was probably naive to think this momentum would have lasted, but for a while, our efforts mirrored those textbook explanations of how a bill becomes law and how the roles of three branches of government functioned. Today, mistrust of government is widespread, as is suspicion of science. If we believe that we can't trust our elected officials or scientific experts, it creates fertile ground for embracing bizarre theories and rationales based on what my science training describes as the "sample of one," e.g., "They say X is a problem, but nothing is wrong with me!"

With governments engaged in self-loathing from the top, many state and provincial legislative bodies have been actively striving to *prevent* regulations that would safeguard public health and the environment based on the assertion that regulation by itself is a negative thing. The rationale seems to be that regulation, by default, is big government breathing down your neck. Moreover, it might require public money to implement, and how dare we spend public money on something as loathsome as regulation. How did we get here, and what does it mean for future efforts?

In the United States, I suggest we start with Reagan. His 1981 inaugural address declared: "Government is not the solution to our problem. *Government is the problem*." When those seeking to govern assert that government itself is problematic, it invites cognitive dissonance, doesn't it? I thought this March 2021 letter to the *New York Times* editor from author, educator, and activist Mimi Abramovitz nailed it.

> For the last 40 years our top leaders convinced the American public that government is *too* big and *very, very* bad. They sealed the small-government deal with endless tax cuts, spending cuts and privatization of the public sector.
>
> Not surprisingly, according to Pew Research Center, Americans' trust in government plummeted from 73 percent in 1958 (Eisenhower), to 54 percent in 2001 (G.W. Bush) to an all-time low of 17 percent in 2019 (Trump).[1]

If government is very bad, what role does the democratic process play in policymaking? Is policy bad? Is government by the people for the people bad? Are we the people bad? It boggles the mind.

History includes failures in government and leadership, but human institutions are rarely perfect. While some administrations have been downright corrupt, that doesn't mean democratic government itself is bad. And frankly, what's the better alternative? Churchill (quoting an unnamed predecessor) made this statement to Parliament in 1947: "Many forms of Government have been tried, and will be tried in this world of sin and woe. No one pretends that democracy is perfect or all-wise. Indeed *it has been said* that democracy is the worst form of Government except for all those other forms that have been tried from time to time."[2]

I ask, what is democracy without government?

So far, cutting agency budgets, privatization of public functions and market-based strategies have not successfully addressed the PFAS crisis, invasive species, algal blooms, contaminated groundwater, lead water pipes, or climate change. I fail to see how cutting taxes will fix any of this. Imagine Eisenhower trying to build the interstate highway system with a "government is the problem" cloud hanging over him (or with today's marginal tax rates for top earners—less than half the 91 percent level the United States imposed from 1953 to 1961).

As the Reagan philosophy permeated U.S. politics, it changed the environmental tide. By elevating and empowering agency heads like James Watt and Anne Gorsuch Burford, who sought to dismantle environmental protections, Reagan also laid the groundwork for what is now an unfortunate partisan divide on environmental issues. None of the major advances in environmental policy in the 1970s and 1980s would have happened without significant bipartisan support. The irony is that conservation and conservative have parallel roots as words and philosophies, but few people who call themselves conservatives in the national spotlight today seem to embrace that alignment.

During the height of our success, we were not only moving new policy through Congress but also securing authorizations for public money to implement the policy, as well the appropriations to get the money in the bank to do the work. My Sierra Club colleague Larry Williams, who oversaw our lobbying efforts on the federal budget, would sometimes tease the rest of our team by saying, "Well, at least my bills always pass" (even if it was after the drama of last-minute wrangling as the clocked ticked on a closing budget cycle). The combination of strong policy and funding is powerful, but weak policy and scarce resources undermine progress.

Congress has passed only a handful of significant new environmental policies in the last thirty years. Seriously. While many of my colleagues have discussed the need to bring fundamental planks of the environmental law framework (such as the Clean Water Act) up to date with current environmental conditions and scientific knowledge, no one dares bring them to Congress. Opening any law of this import to a divided and marginally functional House and Senate (and/or potentially hostile president) risks losing ground or outright evisceration. Thus, we work within the old framework, relieved that it is still in place, but constrained from making new gains or addressing old gaps. The environmental movement has been in the defensive mode for so long, it is hard to imagine driving new policies.

Advocates have found different pathways to make progress. For the Great Lakes, the biggest game in town since 2005 has been the Great Lakes Restoration Initiative. Following the 1987 GLWQA, the question among advocates and agencies was how to tackle its broad agenda. A 2003 report by the U.S. General Accounting Office (GAO) raised attention across the community. Cumbersomely entitled *An Overall Strategy and Indicators for Measuring Progress Are Needed to Better Achieve Restoration Goals*, it seemed to some as an effort to undermine EPA's authority and its Great Lakes National Program Office. Others saw it as a basic observation on the need to coordinate the multiple agencies and layers of government involved in Great Lakes management. Regardless, in 2004 the George W. Bush administration signed an executive order to establish a multiagency Great Lakes Interagency Task Force, and charged it with producing a $20 billion U.S. action plan to restore the Great Lakes.

In this same period, U.S. Great Lakes groups convened a series of exploratory meetings on framing a new initiative for the Great Lakes to take on lingering challenges and growing threats such as the rapid expansion of invasive species. I welcomed the invitation to participate. A key meeting took place in Grand Rapids at the

Steelcase Corporation's conference center (great chairs!) with leaders from around the region, hosted in part by philanthropist Peter Wege. A central question was: "What would it take to clean up the Great Lakes?" with a focus on the U.S. agenda. Not surprisingly, the participants identified well-known and long-standing needs, from restoring habitat to cleaning up the Areas of Concern, as well as a range of other challenges. In my small group session, a seasoned colleague estimated that taking on the U.S. AOCs alone would cost $20 billion. Another responded: "That makes sense, but Congress will have no stomach for a number that big."

Lake advocates were actively seeking guidance from the region's leading scientists. As I mentioned earlier, in 2005 a team of Great Lakes scientists, led by former SAB members Al Beaton and Henry Regier, issued the paper "Prescription for the Great Lakes: Ecosystem Protection and Restoration (Avoiding the Tipping Point of Irreversible Changes)." This succinct and potent analysis explained how the combined six threats of nutrient loading, toxic chemicals, land use practices, historic overfishing, invasive species, and alterations to the natural hydrology of the Great Lakes system were reducing ecological resilience in ways that could lead to ecosystem collapse. The paper remains one of the most compelling cases for holistic approaches to protecting the Great Lakes, and echoed the landmark conclusions in the Millennium Ecosystem Assessment Synthesis report published that same year: "changes being made in ecosystems are increasing the likelihood of nonlinear changes in ecosystems (including accelerating abrupt, and potentially irreversible changes) with important consequences for human well-being."[3]

Shortly after the "Prescription" publication, the Interagency Task Force released its own preliminary report for public review. Meanwhile, the environmental groups had built a sufficient network to formally organize the new Healing Our Waters Coalition (HOW) which would rise to shape and champion the new federal initiative. In September 2005, HOW held its first annual conference. The common agenda that emerged from these efforts was a short list of six areas where investments in federal funds would make a big difference: 1) controlling invasive species, 2) restoring and protecting fish and wildlife habitat, 3) restoring nearshore/coastal waters through reducing overflows from municipal sewage treatment plants and industrial waste, 4) cleaning up contaminated sediments in the twenty-six U.S. and five binational AOCs, 5) reducing nonpoint pollution from agriculture and other land-based sources, and 6) reducing persistent toxic pollutants through cleanup and prevention. It was a good list, with considerable ambition.

The HOW coalition built on many of the tactics we had developed in the 1980s, including Washington Weeks, with volunteers from the region lobbying their congressional delegations. To their credit, the Healing Our Waters-Great Lakes Coalition built a large alliance around the initiative—the last time I checked, there were 175 member organizations. Since 2010, their efforts have succeeded in securing close to $5 billion for cleanup and restoration projects, including $1 billion from the 2022 Biden administration infrastructure package. But the focus from the beginning was on U.S. dollars spent to fix problems on the U.S. side of the border. While HOW grew, GLU faded, along with the binational community that had grown around GLWQA and the binational agenda that had emerged from that community.

I don't want to take anything away from the initiative's achievements to date, but there are some strategic limitations to a campaign focused on appropriations to repair the damage of previous decades. My colleagues grew tired of me saying "No one marches with a sign that says "appropriate!" My point was that lobbying for dollars for incremental progress is different from lobbying for a compelling policy change that will address root causes or drive transformative progress.

For more than a decade, efforts to replicate the initiative in Canada largely failed. Part of this has to do with the different structures and roles of the federal and provincial governments in Canada, compared to the U.S. federal and state relationship. Another factor was that the U.S. tactic of "we're putting our money on the table, what's wrong with you?" was not a persuasive approach with our neighbors to the north. It was another subtle influence on the fading binational approach to Great Lakes strategies. However, in 2023, Canada committed to allocate an additional $420 million for Great Lakes preservation and restoration over ten years. It's progress.

While the HOW Coalition has succeeded in securing federal dollars for cleanup and restoration, many state and provincial regulatory programs have been under prolonged budget assault. Administrations led by Republican governors, Conservative Premiers, or constrained by Republican-majority legislatures, have been shrinking their regulatory and enforcement capacity, and thus losing institutional expertise and capacity in the cause of starving big, bad government. The quiet and diligent work to review pollution permits and ensure compliance with them, inspect hazardous waste sites, monitor contaminants in groundwater, and maintain countless other safeguards, falls on the shoulders of the state and provincial agencies. This work has increasingly been carried out by fewer people with less institutional experience and fewer resources than in prior decades. Of course, things will fall through the cracks, and agency morale will falter. My husband and I know far too many former Wisconsin Department of Natural Resources staffers who just couldn't stand it anymore and opted for retirement, jobs in other states, or the private sector. Who could blame them?

The legislative anti-regulatory fervor in my home state of Wisconsin is so bad that our Department of Natural Resources (or any other public state agency) can't promulgate rules to protect public health and safety without a convoluted process that can take thirty months or more. Moreover, if a rule's implementation might cost more than $10 million over a two-year period (the offsetting financial benefits are disallowed in this equation) agencies can't move forward without an act of the legislature, whose current majority seems to act as though its primary constitutional duty is to prevent the executive branch of government from doing anything of substance.

We look, at times, with envious eyes to Minnesota, where in 2008, voters established a special fund through a constitutional amendment (the Clean Water, Land and Legacy Amendment).[4] This small but mighty three-eighths of a percent addition to the state's sales tax supports a range of water protection and land conservation projects, as well as the arts, education, parks, and trails. The fund allows something rare in a world with short-term budget cycles—the ability to plan and execute projects that can make a difference. Thank you, voters of Minnesota. It is, however, the only state in the region with this remarkable resource.

And still, funding the cleanups, restoration, and state implementation of environmental enforcement programs doesn't get us to root causes and the next generation of long-term solutions across the region. Finding those new pathways to policy, practice, and stirring the cultural values that can drive transformational change (and rapidly) is our urgent challenge.

My colleague Todd Ambs reminded me of another kind of erosion in our democratic process—the loss of agency capacity (and enthusiasm) for meaningful public input. The public participation process, from testimony at formal hearings, to public comment periods, listening sessions, advisory boards, and other ways for people to engage, isn't easy, but it is certainly important. An agency official called me several years ago, ostensibly seeking my ideas on ways to improve their public input process. I suggested they consult with colleagues at the university extension service who were doing some excellent work on facilitation techniques and meeting design. As our conversation went forward, I realized their notion of improvement was to dissolve a well-established advisory board with a solid reputation because they didn't want to fund the meetings or manage the coordination. My heart sank.

It takes skills and funding to engage people from a broad public spectrum. In public agencies under the gun to shrink government, mechanisms to open the dialogue to the public have shrunk in tandem with cuts in resources and staffing. Starving agencies can't successfully build public dialogue, nor can agency cultures that just don't want to bother with cumbersome public input anymore.

One way we successfully reached the public was through the news. The journalists and editors who paid attention to Great Lakes issues and published investigative reports and potent editorials were essential to our success. I can't imagine winning the Michigan Wilderness campaign without the *Detroit Free Press*'s Hugh McDiarmid or editor Barbara Stanton. Environmental reporters from the region's papers never stopped asking tough questions. They helped set the Great Lakes public agenda, and told the hard stories, from toxic contaminants to threats to the dunes to attempts to intimidate Great Lakes scientists. There are still a handful of outstanding journalists who report about the environment in the region, but they are a rare breed in today's news organizations. The loss of local journalism sources in so many smaller communities, and the consolidation of ownership into a few large media conglomerates, limits access to information and discourse. A canned article or tweet simply can't replace investigative journalism. Going back to Lee Botts's efforts on nuclear power, there were clippings from the *Dixon Evening Telegraph* (Dixon, Illinois), the *Herald-Press* (St. Joseph, Michigan), the *Herald-Palladium* (Benton Harbor, Michigan), and others. Small community newspapers carried weight and shaped public opinion.

Social media is filled with faceless insults, posts peddling conspiracy, and falsehoods cloaked as facts. To screen this out means retreating to self-selected networks that affirm our established beliefs. It doesn't feel like social media (beyond some thoughtful blogs and podcasts) belongs with traditional journalism as the fourth estate of the "power of the realm." I'll admit that social media has the capacity to influence and frame political issues, but it is with reactive breadth and rarely with investigative depth.

In the 2020s, our culture struggles with deeply divided belief systems about the role of government and the workings of democracy. Conspiracy theories have created conspiracy cults, "troll farms" seek to influence political opinion via social media in stealthy and effective ways, and people describe our democracy as broken. Public discourse and mean-spirited political campaigns spew anger, insult, and, too often, a pack of lies. Substance, reason, and logic rarely get the stage.

When we were most successful in the "then" days, we were shaping policy in a system where we took its basic functionality for granted. We centered issue campaigns on public policies to protect the environment and public health. The three branches of government were a means to achieve those ends. Now, the government itself is under attack, while its critics in power use its dysfunction as a tool to realign public policy with increasingly constrained roles for the public. I remember bipartisan floor debates where both sides could make their case and disagree without trying to annihilate each other.

It doesn't mean there weren't power plays, or heated exchanges, or the widespread political practice of "reward your friends and punish your enemies." Today, political strategy seems wrapped around simplistic and destructive practices such as "kill anything the opposite party introduces" as the primary objective. If political strategists have reduced the goal of public office to holding on to power for its own sake, it is logical for voters to deduce that the concept of representative government is at risk.

People talk about tackling gerrymandering and rethinking our approach to primary elections (which tend to play to the political extremes in a party) as ways to return to more even-handed government. I'm all for it. But, to offer a hometown example, when, after a multi-year process, your state Supreme Court summarily imposes legislative maps that favor the party in power for another decade (conveniently on the last day of their session), it erodes your confidence in the fairness of either the legislature or the judiciary. Losing confidence helps fulfill the neoconservative prophecy of government's brokenness, in which they are culpable as the breakers. It is also a warning sign.

My instinct is to resist the brokenness as a done deal and argue that our institutions can survive this assault, but sometimes it is hard to rally. We are living in a period of rapid and significant change with an unknown new era ahead. In a world in which authoritarianism is on the rise, our climate is rapidly changing in dangerous ways, and disparities in wealth and opportunity are stark, billions of people just want a safe and decent life. We have the opportunity to shape a new era that could be transformational in reimagining ways to thrive on Earth. The extreme alternative could be very bleak. With whatever tools we have remaining in our strained democracy, we need to keep striving for the possibility of positive transformation and a dawn after the darkness of these turbulent times. I want to assert that surrender is not an option, but also acknowledge that history shows us that governments, like ecosystems, have tipping points, after which there is no turning back.

CHAPTER TWENTY

Imagine Thriving

If we find that we are facing brokenness, how do we move forward? Knowing that old systems and stabs at solutions have only brought us to a place that isn't good enough, let's imagine other options for a healthier and more vibrant future.

I'm not going to suggest that we all crawl into a cave and eat sticks to reduce our environmental impact. I'm suggesting that we're not helpless and we *can* envision a different way of living with the lakes and our regional environment and take the steps to bring that vision to life.

Sustainability has become something of a buzzword in environmental circles. In the evolution beyond boom-and-bust strategies, it dares to imagine a relationship with nature that sustains essential ecological health, as well as social and economic well-being. In business, this became known as the "triple bottom line." Like many attractive concepts, its meaning and goals have been interpreted and adjusted to suit various endpoints, and just calling something sustainable doesn't mean it is, but it might look good in an annual report. As such, sustainability has lost some of its punch.

I like this expanded view from Daniel Christian Wahl: "*Sustainability* refers to sustaining the underlying pattern of health, resilience, and adaptability that maintain this planet in a condition where life as a whole can flourish."[1] He argues that we shouldn't be striving to sustain the way we currently organize modern society, but rather striving to *redesign* that organization using a regenerative perspective.[2] My colleague Michelle Miller—a skilled systems thinker—offered another dimension of sustainability from the perspective of Indigenous cultures and the concept of "right relations." She points out that successfully coupled human and natural systems depend on "right relationships" between humans and the land. When the relationships are right, health—physical, social, ecological—can emerge. The "rightness" requires respect and reciprocity within and among all of creation.

Resilience, regeneration, and reciprocity are not bounded by a bottom line, unless it is health and well-being. Just to get our terms clear, *ecological resilience* is the ability to rebound from, and the capacity to adapt to, disturbance and disruption—like a forest's capacity to grow back after a fire, or a community's capacity to withstand extreme storms through smart floodplain protection. *Resilience* is a way of enhancing our options to survive and thrive. *Regeneration* is an act of restoration and investing in the capacity to be resilient, such as farming in ways that naturally enrich soil health and prevent erosion or restoring native plants to enhance habitat for species that depend on them. *Reciprocity* is a benevolent mutual dependence or, to quote author Robin Wall Kimmerer, "the give and take with the land . . . living as if your children's future mattered, to take care of the land as if our lives, both material and spiritual, depended on it."[3]

On the way to resilience, regeneration, and reciprocity, we need to identify practices and assumptions that can lift us from exploitation to a new partnership with the environment. First, let's get what we *don't* want out of the way. I don't think we want to live in a degraded ecosystem with rusting industrial towns and gloomy environmentalists telling us it is only going to get worse.

Like seagulls, rats, or even cockroaches, humans can adapt to a wide range of environments, including some extreme or degraded ones, and keep the species going. Good old *Homo sapiens* has invented ways to live in the Arctic cold, the searing deserts, and in urban canyons far from a blade of grass. In degraded environments lacking beauty and variety, we might not live long, healthy lives, but we might survive, as people do in squalid slums and neighborhoods downstream from chemical plants with nicknames such as "cancer alley." But who would wish that this is all we should aspire to, or presume that we have the right to make the choice to impose these conditions on future generations and the rest of life on Earth? I don't know whether a seagull cares whether its meal comes from the landfill garbage or the sparkling sea, but *I* care, and perhaps that's the difference between us and the other scrappy adaptive species of the earth—a conscious awareness of alternatives and a yearning toward the fresh air, clean water, and beauty over the stench of garbage.

Let's also consider what my colleague Jim Ludwig describes as "elegant" species—those with special niches and needs for high environmental quality—species that wouldn't thrive just anywhere on the scraps the world might throw their way. Nature expresses this elegance through eagles, butterflies, trout, trillium, wild rice, leopard frogs, and millions of other wonders. The elegance is not mere aesthetics; it is purposeful. Every unique individual, every species, population, habitat, and ecosystem is actively playing a part in maintaining and perpetuating life on Earth, whether they know it or not. We, of all the species, have the ability to know. Knowledge can be power if we choose to act on behalf of elegance and survival—roses as well as bread.

So, how *do* we want to live? I propose that we envision elegance, beauty, and purposeful lives and livelihoods in a more just and equitable society. It will take bold imagination to conceive how we can build a healthy, nontoxic, sustainable economy while restoring vitality to our land and waters, but the times call for bold imagination.

What might we do differently as those responsible for the magnificent but fragile freshwater heart of the earth? How do we step up to do our part to perpetuate the beauty and ecological integrity of Great Lakes country and the lifeblood of the water planet? While these are big questions, they're necessary.

Sometimes when transformative change seems unfathomable, I remember a talk by Dorothy Lagerroos, one of the coauthors of *The New Science of Sustainability*. During the Q&A, someone asked, "How do we get there [to a sustainable world] from here?"—because the transition just seems so daunting and impossible. Dorothy responded that people living in the Middle Ages didn't know that the Renaissance was just around the corner. Still, there were people with imagination who planted the seeds for that major cultural shift in earlier eras that made that transformation possible.

Transformation toward resilience will require

- embracing a new relationship with the natural world;
- rethinking the way we govern and "manage" the environment;
- diversifying the region's economy, and embracing "green" farming and manufacturing; and
- advancing justice and equity.

Let us be seed planters, cultivate promising ideas, and make room for what is possible.

One clear lesson from our extractive past is that we must design *with* nature, not against it, which means a new relationship with the natural world. My friend Marian Farrior has taught me about ways we can emulate nature's design in human organizations and systems for mutually beneficial outcomes. Consider, for example, what nature teaches us about monocultures—systems artificially dominated by a single species, like a vast cornfield or a pine plantation. Nature doesn't favor them because they lack the dynamic variety of countless intertwined relationships that keep energy, nutrients, and vitality pulsing through the system. To keep overly simplified systems going, large-scale monoculture crop systems require significant inputs, such as water (often in the form of irrigation), nutrients (fertilizer), pest control (pesticides), and capital (for labor-saving machinery). All these inputs replace natural ecosystem functions with external resources and energy.

Nature loves diversity. In place of a cornfield, nature grows grass-lands, prairies, savannas, and even forests—systems that are teaming with relationships above and below the soil. A healthy prairie makes its own nutrients, creates its own energy, and fends off pests through multiple natural processes. Organic farmers know these principles and incorporate them into their practices.

Food and agriculture offer major opportunities for new relationships. We can't continue current farming practices and expect clean water and healthy soils. Conventional farming practices have helped create vast dead zones—areas starved of oxygen and often laden with silt and chemicals—in water bodies from Lake Erie to the Gulf of Mexico. A hungry world and a healthy planet cannot afford dead zones and dwindling soils perpetually hooked on artificial chemical infusions. The world's dead zones (and there are more and more of them) will continue to expand and render more productive aquatic and marine habitat useless unless we change the agricultural practices that pour nutrients, pesticides, bacteria, and silt into rivers, lakes, and streams every time it rains.

The United States cast its relationship with farming in a new mold in the 1970s when Secretary of Agriculture Earl Butz transformed American farm policy to drive the "bigger is better" model. Since Butz, our federal farm policies have embraced the "produce more; sell abroad" strategy. This has driven small farmers and those who don't produce global commodities into harder and harder times and left the big operations vulnerable to swings in global markets.

Climate scientist Jonathan Foley asserts that agriculture has become an economic trade system, not a *food* system. Scaling up for global trade has driven Midwestern farming to massive soybean and corn fields—crops that aren't (for the most part) producing direct food for people unless you really love your high-fructose corn syrup. For those who cling to the "we feed the world" justification, please note that about 40 percent of the U.S. corn crop goes to make ethanol fuel, which is a) not edible, and b) has its own troubles in terms of "energy in vs. energy out," as well as water quality impacts. We export much of the rest of the corn crop for animal feed, and process only a small portion into products that we eat.

Environmental consequences from large-scale row-cropping include soil erosion, and fertilizer, manure, and pesticide residues flushed into waterways. Those same pollutants can seep into groundwater, as the region's extensive (but not uniform) contamination by nitrates from fertilizer and bacteria (such as *E. coli* from manure) illustrates.

Speaking of manure, a parallel scale-up has transformed animal agriculture. The family-owned farm of our cultural imagination with red barns, happy Holsteins in the pasture, and chickens in the yard is disappearing. Wisconsin alone has lost forty thousand dairy farms over the last four decades, with only about 6,500 remaining in early 2022.[4] As they shut down, the tillable land that stays in farming often goes to the bigger operations and into corn and soybean production, and "Big Dairy" keeps producing milk, especially for the global market.

Much of today's dairy is produced in concentrated animal feeding operations (CAFOs) that can house thousands of dairy cows at one site. (Large-scale hog and poultry producers use similar operations.) These massive facilities—sometimes called factory farms—produce enormous amounts of dairy products and meat, as well as animal waste, in a concentrated area. Dairy barn operators typically wash the waste out of the barns with water, creating a liquid version of the manure. All CAFO operators must obtain a CWA permit to discharge any contaminant into local waters. Some operations store it in open air waste lagoons to break it down with anerobic bacteria and then spray it on fields. What could possibly go wrong? Among other things, if the facility sits on porous geological substrates, as is the case in the karst region of northeast Wisconsin, a leaky lagoon can contaminate groundwater. Heavy rains can flood lagoons and send the concentrated waste into local watersheds.

The more sophisticated CAFOs contain and break down the waste (also with bacteria) in closed digesters that capture the methane in the process. The methane produced by manure management accounts for about 10 percent of U.S. methane emissions, and methane is a potent greenhouse gas.[5] Thus manure management is not just a water quality challenge but a climate change challenge.

In my household, we make the extra effort to purchase our meat and dairy products from farmers and farm cooperatives that work with their land's natural carrying capacity for incorporating animal waste into healthy soils. We can do this because we have access to outstanding suppliers at our local farmers market and a few grocers and specialty shops that put locally sourced meat and dairy on their shelves. I know this option is a privilege. I also know it takes time, personal energy, and a bit more money to do this than just grabbing a random package at the grocery store. If I am going to eat animal-sourced food, I am committed to lowering its environmental impact and supporting the farmers who work so hard to do the same. And, as important as it is to make these personal choices, I know the system itself needs to change.

The Great Lakes region could choose to rise as a world leader in ecologically designed agriculture, or "agroecology," and kick our dependency on the corn-and-soybean monoculture habit along the way. Already, thousands of farms have made that transition, and the body of knowledge and expertise is growing in the field (literally and figuratively), often a product of collaboration between farmers and researchers. If Earl Butz could transform agriculture, why can't a new generation of leaders?

In addition to diversity, nature has another important lesson to teach humans about environmental management strategies. Natural systems are complex and dynamic, and we need to think like a system, manage like a system, understand that we are *part* of systems. David Suzuki sums it up nicely. "When we forget that we are embedded in the natural world, we also forget that what we do to our surroundings we are doing to ourselves."

Current environmental management practices are artifacts of the well-intended but piecemeal approach embedded in laws enacted in the 1970s and structural decisions that put parks in one agency, wildlife in another, pollution permits in in another, and so forth. We don't have an overarching Department of the Environment. Multiple agencies manage overlapping pieces of the pie for different objectives all at the same time. Nature, in fragments, fails to thrive. Only integrated systems, with connections to other systems, function well. The fragments of our human management practices don't add up to a well-functioning whole. Instead, agency and departmental silos carve up goals and strategies and isolate budgets, skill sets, and knowledge. It shackles progress. We need twenty-first-century governance to address twenty-first-century environmental problems. What might this mean?

We could start with ecosystem-based management as an overarching strategy for the environment, and aspects of public health unified by a common goal. By ecosystem-based, I mean designing our management strategies to keep the whole system humming, from predators to decomposers, from headwaters to the oceans, from inputs of energy, nutrients, carbon, water, and life forms to outputs of the same, striving for optimal health and resilience at the levels of communities, populations, species, and individuals. That's what nature's been trying to do with remarkable results for a few eons or so, and then we came along and tried to manage it like the production of widgets.

Any human-designed system wouldn't be perfect, caught in the nexus between democracy, capitalism, and nature, but surely a more ecosystemic approach would be a step in the right direction. For a new approach to emerge, we need to collaborate, seek natural and human feedback, learn, and take a few risks.

In 2019 and 2020, I served as an advisor to a project led by the River Alliance of Wisconsin. They organized it around a bold challenge. If we weren't constrained by the existing environmental policy framework, how would we design a framework that *could* achieve and sustain the goal of clean, abundant water? This was no mere academic exercise—more a gauntlet thrown down before a diverse set of leaders who had all wrestled with some aspect of water policy and management over decades. We agreed that the status quo isn't working. Our group included several environmental leaders, former and current agency staff, two small-business owners (one of whom also farmed), two former tribal chairmen, a community organizer, two scientists, and others with unique expertise.

Over a series of meetings, we wrestled with the topic. Hard and honest conversations included self-critique of the accomplishments and failures of the last thirty years, and a frank discussion on the role of colonialism and racism in shaping environmental policy and practice in North America. At the conclusion, we didn't have volumes of new legislative proposals to offer, but we did have what the group called "The Water Agenda"—twelve succinct principles, which, if applied in concert, characterize a system that could achieve health and ecological goals for water quality, water supply, and aquatic habitat. While we developed the agenda with state-level management in mind, its twelve principles can apply to multiple scales. Here's a shorthand version: a new system for managing our water must

1. Establish a single, enforceable, common goal to align public and private action to ensure abundant water, in perpetuity, that supports and sustains the region's ecosystems—water that is accessible, affordable, and available for all people, plants, and wildlife without risk.
2. Strengthen the public value for water (as part of our culture, heritage, and identity).
3. Engage all stakeholders; embrace inclusive, equitable, fair, and comprehensive decision-making.
4. Manage water as it exists in nature (the water cycle, watersheds, connected surface and groundwater).
5. Prevent threats to people, plants, and wildlife; embrace the precautionary principle (test chemicals before use in products and processes).
6. Connect benefits and harm (so the public doesn't bear the full harm and risks imposed by private producers).
7. Emphasize education about water for all ages. We all need to know more about water to make sound health and environmental decisions.
8. Focus on the causes of a problem, not the symptoms. Quick fixes are not sustainable.
9. Regenerate and restore. Sustaining degraded conditions is neither sufficient nor acceptable.
10. Be sufficiently funded. If want a new system to work, it needs stable and sufficient funding.
11. Adapt and be flexible. Like nature, systems must be dynamic and capable of changing to meet new needs.

12. Generate knowledge to solve problems. Keep learning, anticipate, and look ahead.

Is this all just pie in the sky, or is it a reality check? None of us were advocating that we just toss out the current regulatory framework, but if we're going to get unstuck, we must imagine what *could* be and begin to move in that direction. Systems tend to reinforce themselves unless we change them, and someone must lead the change. Let's continue to reimagine systems that advance resilience, reciprocity, and a healthy democratic process.

One of those systems needs to be our regional economy. In 2006, I spoke with John Austin, author of the Brookings Institution *Vital Center* report that explored this topic. He explained that a lot of twentieth-century economic thinking was embedded in two waves following European colonization. In the settler era, prosperity came from extracting natural resources and cultivating natural goods—often for distant markets. Lumber, fish, livestock, grain, and other products helped build the towns, ports, and rudiments of the modern transportation system from which the manufacturing era rose in the twentieth century. In the industrial wave, we pulled raw materials from the earth and manufactured products at a scale unimaginable to those early settlers, thanks to innovations like the assembly line and the cheap energy of the times. Both were variations on extraction-based economies.

It seems neither wave gained much wisdom from the people who were already here prior to those eras, including practices that sustained food supplies, supported extensive trade networks, and maintained clean water and healthy lands. Early in her career, Michelle Miller studied with Keewaydinoquay Pakawakuk Peschel, an Anishinaabeg elder, teacher, and ethnobotanist. Michelle shared that Keewaydinoquay explained that one of the Indigenous peoples' important practices is paying attention to what we might call a high level of "feedback" from the intertwined human and natural systems—noticing change, learning, adapting, and understanding the dynamic balance in these systems. Another fundamental principle is that the earth abundantly provides. When systems become imbalanced—such as food scarcity—it is a signal that humans must change their interaction with the system to restore balance.

In contrast, beliefs that we must bend nature to meet human needs, regardless of the cost to natural and human systems, influenced the settler and industrial approaches. Many of the daunting challenges we face in the Great Lakes today are because we didn't listen to or look for feedback from the natural system (or the economic system) before problems grew to their current scale.

In 2020, the Chicago Council on Global Affairs published an update to *Vital Center* in *A Vital Midwest: The Path to a New Prosperity*. In it, the authors assert:

> the Midwest is neither an economic monolith nor, despite lingering popular misconceptions, a "Rust Belt." Many Midwest communities, large and small, have successfully evolved from their industrial and farming roots and are winning in today's globalized, tech-based, and knowledge-driven economy.[6]

It also acknowledges:

> Today there are two Midwests. While many communities have successfully outgrown their industrial shells and boast diversified and dynamic economies, the region is also home to the country's largest number of older industrial communities that, having lost anchor employers, remain adrift.[7]

So, we're shaking off the Rust Belt image, but many communities are still struggling, and we carry the lingering burden of those early economic waves with us. That includes the concept of "natural resources" ready for the taking with little regard for the consequences of the "take," and that's how we built material wealth. After a few centuries of cultural reinforcement of the "resources" frame, challenging its assumptions is hard. However, nature, despite its bounty, has limits, as does an economy based on finite resources.

In a meeting long ago, I inadvertently offended someone when I stated that capitalism was an amoral system. He thought I had said "immoral," and he had risen with some umbrage to refute my assertion. My point was that capitalism doesn't come with a moral compass. It is an economic system based on the accumulation of capital. Any functioning society has moral and ethical values and standards that shape its economic and social systems. Thus, we have things like child-labor laws, meat inspectors, and, yes, environmental laws and regulations because they provide societal protections and safeguards that we value. Do they place constraints on free-market capitalism? Of course. The safeguards have higher social value than the cost of those constraints. (This worked better when there was greater cultural agreement that one of the roles of government is to provide some level of safety and security for the people, but I caution against abandoning the principle in rocky times.)

My other quibble with capitalism is the long-established practice of passing along the cost of corporate "externalities" to the public to absorb without the consent of the affected public. We have absorbed the storage of industrial chemicals in our air, water, and tissues, the elevated levels of childhood asthma in urban centers, and the obligatory and unknown costs of safely storing radioactive waste for thousands of years. If corporations had to cover the true costs of doing business, they would have to internalize these externalities, but they don't—meaning that we do.

What could happen if we reimagined our economic relationship with nature? Can we step away from the "resource" framework based on extracting finite raw materials from the earth, depleting soils, and industrializing once-vibrant forests? What if our approach was less about "taking from" the land and waters and more about our active partnership and relationship *with* the rest of nature? Can we embrace reciprocity? If our economic activity shared a commitment to sustaining and regenerating the region's natural assets—its clean water, rich soils, remarkable beauty, and other amenities of this water-rich landscape, we and the natural systems would both benefit.

For a region with a deep history in manufacturing, why couldn't the Great Lakes region become a global center for technologies that serve the needs of humankind and nature? We could build the nation's first regional high-speed rail network. Elsewhere on the carbon reduction and energy front, we could step up to design, engineer, and build renewable energy technology and improve energy efficiency in buildings, manufacturing, and transportation systems in our cold-weather climate. Thinking beyond fossil fuels and the internal combustion engine isn't a burden. It is a shining opportunity for innovation and leadership. Solar and wind technologies have already made rapid gains in the region, and jobs in these sectors are far outpacing jobs in conventional energy. In 2020, despite declines due to the pandemic across sectors, "more Midwesterners worked in clean energy than worked as accountants, auditors, computer programmers, web developers, and real estate agents and brokers combined."[8]

In 2021, Tonya Allen, president of the McKnight Foundation, raised this question in a virtual talk (and I'm paraphrasing here): "If we can't solve the challenges related to climate change in the American Midwest—a center of energy production, industry, agriculture, and a region with remarkable scientific and technological capacities—how can we expect anyone else to succeed?" I sat with that assertion for a while. She's right. *Will* we do it?

A crosscutting challenge is building equity and justice into all our systems. Healthy communities and economies depend upon healthy and high-functioning ecological systems. If we want to thrive, we need to protect the natural functions and systems that produce clean air and water, and those that keep soils alive and fertile, as well as those that support vital natural functions like pollination, decomposition, and carbon sequestration, among others. This is the true "hand that feeds us," and it is foolish to abuse it or take it for granted.

Healthy communities also depend on healthy people, from physical health to things that foster mental health and stability, such as security, safety, the absence of routine trauma, and a sense of belonging. Reducing the disparities between thriving communities and struggling ones (which *A Vital Resource* rebrands as "communities in recovery") is more than just overcoming shifting economic winds in the region. The health and resilience of our human communities requires social justice and equity, meaningful inclusion, and representation for all who seek to participate in public decision-making.

It also requires access to healthy food, safe shelter, good health care, and opportunity. Stark inequities lead to conflict, injustice, environmental degradation, and despair, as Flint and other communities have taught us.

Social justice and equitable access to resources and goods foster health and successful human relationships. When more people have a decent share of the pie, *and* the pie gets bigger (and I'm not just talking about material wealth here), it reduces conflict and builds community. Functioning human communities are better able to protect the environment; those in crisis cannot. Likewise, degraded environments, burdened with legacy pollution and failing infrastructure, cannot support resilience in human communities.

We have many hurdles to surmount on the journey to greater justice and equity. When a society concentrates poverty in urban centers with aging infrastructure, and when we can't provide basic social needs such as safe drinking water or good schools in all our communities, then we're not functioning as a robust or sustainable society. We're going to have to wrestle with racism and xenophobia and their long legacies to build and rebuild more vital human communities again in the once-great cities of the Great Lakes, as well as small cities and rural areas. This is hard, deep work, but in a region that claims to be good at hard work, surely we can rise to this task.

When we look at communities with stronger vital signs, we can see common factors such as a diverse economic base as well as investments in natural assets such as lakefronts, riverfronts, and parks. They embrace, and celebrate, human creativity and diversity, from the arts to food, culture, and history, and an intangible but firm sense of pride that people associate with "their place." They also have practical systems such as mass transit that provides many people mobility with or without cars, access to health-care services, housing options, and access to broadband. These assets help support the people who in turn bring life and energy to the system, and who weave the fabric of vibrant communities where people want to be.

The relationships between robust economies, healthy environments, and strong human communities aren't linear; they feed into each other. There's nothing resilient about exhausting resources, people, or places. We could choose to let go of that old pattern in the Great Lakes region (it appears to be abandoning us anyway) and fully emerge from the Rust Belt. We could be the Water Belt, the interior coast, or that "vital center" that the Brookings Institution envisioned.

Transitions can be emergence or upheaval. Can we envision a humane transition instead of one that leaves broken lives in its wake? If so, it requires moving beyond fear and creating bridges so that people in the transition don't fall through the cracks.

Let's envision abundance—a healthy, productive Great Lakes region that sustains itself, filled with natural beauty and creative and energetic human communities. Imagine prosperity and community, built on strength, knowledge, and perhaps a few leaps of faith, sustained by the natural gifts that could give again if we rise to the challenge of restoring and protecting the living system we call the Great Lakes.

CHAPTER TWENTY-ONE

Wonder, Love, Grit, and Wisdom

If we frame a vision, conceptualize the big steps, and begin to act, we're on our way to forging solutions. But the "change-maker" toolbox has other capacities to help a society transitioning to health and resilience. Four essentials are: wonder, love, grit, and wisdom.

Wonder

Wonder seems in short supply these days, and humans *need* wonder. That's why people try to sell it to us in packages like a trip to Disney World, or the newest incredible-graphics video game. But there's something hollow about artificial wonder—something that leaves us wanting the new-and-improved version. Wonder 2.0?

The real thing is substantial. It feeds the soul, inspires the imagination, humbles human bravado. The Great Lakes are packed with wonder. Their vast blue expanse boggles the mind as the eye looks for what might be beyond the horizon. And what is beyond the horizon?

More water. We're talking about the stuff of life—the astonishing bond between two gases that produces a crystal-clear liquid that flows in vast quantities only here on Earth—*out of all the known cosmos*, only here on Earth. While astronomers have identified vapor or other signs of water on exoplanets light years away from us, where life depends on liquid surface water, Earth still seems to be the best option. That alone should leave our heads spinning. Why this little planet? Why this solar system? Why this precious blue and green sphere of life?

If contemplating the cosmos overwhelms us, we can explore wonder at a smaller scale. Let's consider the moods of lakes, big and small, ever-changing in design and display, mysterious and unpredictable. The same space might be blue and placid under a summer sun, filled with skipping white caps on a windy day, whipped into foaming frenzy in a storm, brooding in battleship gray under darkening clouds, polished pewter under winter ice, heaped into frozen ridges or a white plain of silence in the dead of winter, or dotted with cloud shadows chasing over the breadth of pulsing blue open space viewed from a window in the sky. A lake can clatter and groan with spring breakup in an onshore wind, lap quietly in lolling ripples on lazy day, crash and roar with waves and wind pounding on a stormy shore, and slip into the exquisite quiet of still water at twilight, brushed with the faint hum of insects in the humid air.

The life that populates this corner of the world offers countless gifts of delight and amazement. Consider the flash of cardinal wing, goldfinch, or bunting; the great flocks of spring and fall; the call of goose and loon. Ponder the act of migration itself. Listen to the chorus of spring marshes or the sound of autumn coming when crickets begin to chirp. Marvel at the glint of silvered fish scales in the sun, a glimpse of an otter peeking out of a quiet cove. Thrill to the possibility of hearing wolf voices. Pause to appreciate the engineering feats of beavers or the longevity of turtles and sturgeon.

Inhale the scent of balsam, warm pine needles, and wintergreen on a northern shore. Appreciate, profoundly, the wind-sculpted bonsai form of white pines. Stand surrounded by a birch grove under a blue sky and drink in the dazzling patterns of blue, green, and white. Savor the glow of autumn in the late afternoon, the tender grays of winter woods, snow itself. Celebrate buds of spring in delicate green, silvered white, or gaudy chartreuse. Be astonished by dwarf lake iris, lady slippers, horsetail grass, or a maple forest carpeted with white trillium.

Under it all, and breaking through the surface in cliffs and ridges, are the bones of the continent. Accept the invitation to be awed by ancient rocks scoured and shaped by ice, water, wind, and gravity. Be amazed by cliffs and turrets, boulders, shelves, cobbles, pebbles, and the fine sand—underfoot on a damp shoreline or heaped in shape-shifting dunes. Admire glacial soils rich with minerals finely ground by miles of ice, packed with nutrients from prairie grasses, the leaves of forests, and alive with the myriad micro-workers of the original world of recycling.

The stuff of wonder is all around us.

Somehow, we must help those who don't know the wonder of Great Lakes country to discover it, and those who do, to discover it again and again until we are reminded that we are bound to act as guardians of the wonder, so that others will have this gift. Surely this is one of the things that mark what makes a people great. This is an act of civilization.

Whether it is as simple as taking a child to a place where they can discover nature, or as ambitious as multiplying the effects by creating more places like the Dunes Learning Center, or just unplugging ourselves every now and then from the electronic barrage that seeks to fill our days, we need to create opportunities to cultivate wonder. Wonder is the original form of "out-of-the-box" thinking. It invites curiosity and expands our horizons, and we need that capacity to envision and realize strategies for a healthy, vibrant world in which the Great Lakes will be a shining jewel.

Love

What's love got to do with it? Just about everything. I love the lakes, the water, the forests, and their gifts, and I am unashamed to say so. Countless others share this love, but many have a hard time expressing it. It's difficult for us to talk about love in this culture—at least the real thing. Midwestern stoicism will bear up through hardship, winter, and deprivation, and not show emotion. Emotion might indicate weakness, and the weak will not survive.

Dave Dempsey and I were reminiscing about the hurdles we crossed to pass the Michigan Wilderness bill. He told me a story about a meeting he attended with Governor Blanchard to secure Tom Washington's support for the legislation. As I mentioned earlier, wilderness just wasn't Tom's thing, but Dave's story shed new light on why. At one point in their conversation, Tom said something along these lines: "Do you know why Jane Elder wants to save these places? She says she finds *peace* there. For God's sake!" Dave reports that Tom's disgust was palpable. Peace, beauty, and love for wildness were clearly not justifications for land management decisions. Even raising these topics was embarrassing.

It made me feel better at one level to know that even the governor couldn't bring Tom around, but wistful about a huge gulf between two people who really cared about the land and waters of Michigan. Then I told Dave my own story. In 1985, I was invited to testify at the House hearing on the Michigan Wilderness bill in Washington, D.C. Eight years into the battle, this was my ten-minute window to make the case in front of the committee that would make or break the legislation.

I had written and rehearsed my testimony to make sure I wouldn't exceed the allotted time. Still, few things prepare you for sitting down in front of a committee in the formal hearing room among the flags and the microphones in the heart of the House Office Buildings on Capitol Hill. I made it through most of my testimony, but as I was reaching the final paragraphs, my emotions began to rise, along with the lump in my throat. "Please," I thought, "don't let me lose it now." But I was over the brink, and with little tears leaking out the corners of my eyes, I read the last paragraph with a wavering voice.

> Throughout my ten years of active involvement in environmental issues, I can say that there is no issue that is closer to my heart than Michigan Wilderness. These are places I know and love, places where

> I have known peace. My dream is that others will know the gifts of these places, that they too will be touched by the grandeur of the remnants of a once magnificent forest, that many years from now, others will know wilderness because in our day, people of vision had the courage to act.

I finished my statement and looked up, red-faced, to see the chairman, Congressman John Seiberling of Ohio, dabbing at his eyes too. He then gave me a nod and a warm smile. We each cleared our throats, and the hearing went forward.

In 2006, one of the pollsters consulting for the Biodiversity Project reviewed all the available public opinion research on the Great Lakes. It affirmed broad public support for protecting the lakes, but in his analysis he observed, "People say they love the Great Lakes, but they act more like they *like* them." It made me realize that it is easy for people to admire beautiful shores and sparkling waters, and appreciate what the lakes do for the region's economy. Love, on the other hand, demands more. It is not a passive relationship.

To say we love the Great Lakes is not a casual statement; it is a commitment. Love is hard work—sometimes heartbreaking—sometimes demanding difficult choices, but so powerful. It is also the most profoundly human power in our lives to achieve what reason says is beyond hope. And so much, given our "lay waste and leave wounds" history, seems to argue against hope.

If you love a place, you will not destroy it. You will learn its wonders and learn to seek the synergy between its gifts and needs and your own. Love is an act of deep reciprocity. Human life requires fresh water, and a relationship built on love and respect for this essential ingredient—so much of which lies cupped in a living landscape in a corner of North America—seems like a mutually beneficial arrangement. It might also bring us joy and a bit of peace.

Philosopher and author Kathleen Dean Moore writes compellingly about the need for a love-based ethic for how we can relate to the environment. In her essay "What It Means to Love a Place," she writes: "To love a place is to care for it, to keep it healthy, to attend to its needs as if they were my own, because they are my own. Responsibility grows from love. It is the natural shape of caring."[1] Sign me up, sister. The lakes need a lot of this kind of love.

Grit and Wisdom

I've married these two topics because the more I dug into grit, the more I found that those with determination, tenacity, and resilience often have another quality that boiled down to wisdom.

Every now and then, I get weary. I start thinking about the myth of Sisyphus—that fellow the Greek gods condemned to pushing a boulder up a mountain only to have it roll back down again over and over. When I get down into a Sisyphus mood, it's time for a reality check and some fresh air. To kick the Sisyphus blues, it helps to know that while his efforts were futile, ours often succeed, even if it is just to wear the boulder down a little. We are also not alone, nor are we condemned. There will always be a need for work to protect the life-giving stuff of the planet, but it can be rewarding, even if the climb is steep.

I'm inspired by the people who keep pushing uphill and those who find reward in their efforts. Albert Camus wrote, "The struggle itself toward the heights is enough to fill a man's heart. One must imagine Sisyphus happy." Well, maybe. I'd still like to see a few boulders get over the top. In the early 2000s, I talked with some of the long-haul activists I admire about how they did it.

Over the years that I knew Lee Botts, I'd seen her retreat into deep funks about Great Lakes issues, and she reserved the right to do so. The thing is, she didn't *stay* there. While she liked to talk about strategy, she rarely liked to talk about herself, but in one of our conversations at the "pond house," I persuaded her to talk about the art of activism and social change itself. Ingredients in her recipe for success emerged over the course of our conversation. A common thread in her work and the advice she gave to others is to be opportunistic and observant. Lee had a nose for opportunity. She made the case that you never know when one opportunity—even if it looks like a door closing—will lead to something else down the road.

It might take years or decades for the circle to close, but often new opportunities arise from seeds planted long ago. You can't approach this passively, waiting for pennies to rain from heaven; you need to keep an eye out for opportunities and grab them when they come along. This is a kind of sixth sense—some would call it political instinct. But it's more than politics. It's about a keenly honed awareness that can anticipate that the tide of the times is about to change, or sense who is positioned to leverage leadership at the right moment. I heard someone describe this as being a "horizon watcher."

Even if you can't read all the tea leaves of social change, you can bank relationships and information, which down the road might just create the opportunity you need in a place you least expect it. As I move through the decades, I can affirm that the "relationship bank" has been one of my most valuable resources in trying to make progress. Lee and I talked about how relationships can shift attitudes and practices, and the art and craft of creating the time and space that can form them. She pointed out that you need to create the forums and the dialogue to build human relationships. Dialogue in the right place and the right time can build the foundations for trust, cooperation, and shared approaches to solving problems. Different interests don't just coalesce by happenstance. Someone calls the meeting, extends the invitation, and creates the space where these interests can come safely together. She also advised against assumptions that those with different interests are enemies, and instead encouraged finding common ground when you can. Working *with* can be a lot more rewarding than working *against*.

Community involvement was high on her list as well. An involved community can help shape its future and "own" its sense of place. It might even rise to claim its love for that place. Involvement and action reduce the likelihood of becoming a victim of decisions made in distant boardrooms or legislative chambers. But one's sense of community needs to be broad enough to begin to grasp the larger network of other communities with whom your fate is linked. This leads to the next ingredient in Lee's recipe for successful environmental change: education. Education creates a wider and deeper pool of those who can participate in decisions and shape life at the community level and beyond. Education about our relationships to places and the natural world begins to weave the fibers of informed decision-making that will foster not just survival but abundance.

Edith Chase was another one of the great women of the Great Lakes who dedicated much of her life to educating people about how to make a difference for years. Like Lee, Edith was committed to Great Lakes work and was one of Lake Erie's best friends. She was part of a cadre of tough-minded and warm-hearted Cleveland women who banded together to make a difference through organizations like the League of Women Voters and the Sierra Club.

One of my early Sierra Club mentors, Ellen Knox, was part of that circle, along with Emeline Clausen, a musician and teacher who filled her living room with sheet music and environmental impact statements. They were there, in the heart of the Cuyahoga River Basin, for the first and second waves of Great Lakes activism. Edith advised, "You have to have a little staying power." She also said her longevity in the trenches of activism was due in part to finding ways to have fun, making friends in the work, and taking time to reflect on progress over the years. "The fact that you can't brag about something that happened this month or the next isn't what's important. We *have* made progress." She not only knew the way the policy game is played, but also understood its nuances, flaws, and limitations. People like Edith, who care about, think about, and know about the workings of their government, give me hope for the democracy. When I mentioned this to her, she smiled and said, "We've got to pass it on."

Charlotte Read was another lifelong, tireless leader, based in Indiana's Save the Dunes Council. She observed that first you have to care about a place, and if you care, then learn what you can do to act on that caring. Her suggestion? "Get involved. Join an organization, talk to your friends, your legislator, your senators and congressmen, and other organizations." When we talked, the Indiana Dunes National Lakeshore was celebrating its fortieth anniversary as a unit of the National Park System. While Save the Dunes Council was proudly joining the celebrations, Charlotte affirmed that the work wasn't finished. "There are still dunes, wetlands, and lakefront that need to be protected for public use and protected against abuse."

The world lost Edith in 2017, and Charlotte and Lee in 2019. Each of them remained active in issues they cared about into their tenth decades. These great women of the Great Lakes knew how to get things done, and they opened doors for the rest of us to grasp the torch and keep going.

Indigenous people have offered another kind of wisdom to the Great Lakes community for many decades. Traditional knowledge about the Great Lakes region has finally begun to gain more awareness and respect. Our Indigenous brothers and sisters, whose heritage and experience in the Great Lakes region goes back for thousands of years, have offered to share their wisdom and that of the ancestors. This is no slight thing.

In the fall of 2003, I had the great privilege of meeting Oren Lyons. Oren is a wisdom keeper, or faithkeeper, of the Turtle Clan of the Onondaga people and is known in global circles for his work on Indigenous rights. We were both participants in a meeting called Artful Advocacy, held at the Pocantico conference center in New York, where writers, documentary producers, and others were exploring creative and powerful ways to communicate about serious issues.

In this crowd of heavy hitters, whenever Oren spoke, side conversations fell silent, and he had our rapt attention. He speaks with quiet authority and clarity and is a deep and serious thinker. At the same time, he exudes great warmth and has a wry sense of humor. His incisive wit surfaced during a break when we toured the historic Rockefeller home just uphill from the conference site. After viewing the impressive collections of Chinese pottery from various dynasties and other art treasures in the house, he simply said, with a knowing expression on his face, "Good plunder." During another break, he told stories about a recent gathering of chiefs and elders from throughout North America held in the homelands of the Haida along Canada's Pacific Coast. They had all gone salmon fishing, and he showed us photos of laughing men, including him, dancing in a circle of celebration and thanksgiving with huge salmon in their arms, giving these spectacular fish their respectful due before they became dinner. Earlier, during our meeting, there had been conversations about overfishing and other issues related to biodiversity loss.

Acknowledging this as we looked through the photos, Oren said, "The salmon are willing to share themselves with us, as long as we aren't greedy." It is such a simple lesson. It seems so hard for our culture to put this into practice.

For many years, Henry Lickers was a regular participant at IJC and other Great Lakes meetings. Henry is a Haudenosaunee citizen of the Seneca Nation, Turtle Clan, and was the director and environmental science officer of the Mohawk Council of Akwesasne for forty-three years. Often, the IJC and other groups would tap Henry to play a role in the closing or opening sessions, and he did so with humor and grace. In 2019, Canada appointed him to serve as one of Canada's IJC commissioners—the first Indigenous representative from either nation. At one conference, he offered a call-and-response prayer—a traditional Haudenosaunee prayer of greeting and thanksgiving—and invited the attendees to join in on a spoken response at the end of each verse.

This is a lengthy and detailed prayer, extending to fifteen minutes or more, and some of the assembled stirred in their seats after a bit. I was fascinated at the complexity and beauty of the prayer, which recognizes the great diversity of nature, celebrates its existence, and seeks a statement that acknowledges its value from all participants. In an oral tradition, versions will vary. I found one written version of the prayer where the group response at the end of each passage is "We are all of one mind." Another version concluded each statement with "We now bring our minds together in thanksgiving for . . ." and each aspect is named.

When Henry offered it, my recollection is that he used a question-and-answer process to establish that we were all of one mind. He would speak a statement of thanksgiving and ask, "Can we all agree?" and the group of several hundred participants answered, "Yes, we can agree." Getting a crowd of Great Lakes advocates to publicly state that they agree on one thing is amazing. Agreeing on dozens of principles and observations was nothing short of miraculous. Except, of course, these are things we all know in our hearts in the first place.

For example, after opening the prayer by expressing thanks for the safe arrival of those assembled, the prayer next thanks the Creator and then the waters of the earth, as this version I found on the Ecozoic Times website illustrates:

> Now as one mind we turn our thoughts to the Waters of the Earth for they too have not forgotten their instructions from the Creator of Life. The Waters continue to flow beneath the ground, in little streams and in rivers, in lakes and in wetlands, and in the great seas. They quench our thirst and help keep us clean so we can fulfill our duty to Creation. We now bring our minds together in Thanksgiving to all the Waters of the Earth.[2]

The leader of the prayer then asks for an affirmation of agreement, and after those present respond, the leader states the affirmation before moving on to the next stanza. For example: "We now bring our minds together in Thanksgiving to all the Waters of the Earth."[3]

As the verses continue, they acknowledge purpose and give thanks to all forms of life that live in water, then plants, animal life, trees, birds, the four winds, thunder and lightning, the sun, the moon, the stars, the four spirit beings who live in the four directions, the Creator, future generations, and all nations of the world, ending with "Now our minds are one." The invitation and admonition to find agreement and acknowledge gratitude creates commitment and common ground. No wonder it has lived for centuries as a prayer to bring people together for important work.

I am grateful to each of these leaders for their wisdom and their commitment. They were and are the kind of people who show up and participate. It takes conviction and a certain amount of grit to withstand the opposition, the process, the losses, and to carve out the time and resources it takes to be present. I fully understand that participation, although a right, is sometimes a privilege. If you are dealing with basic survival issues or a community in crisis, your grit is used up just getting through a day. This invites greater responsibility among those of us who do have the capacity to show up to participate with wisdom, tenacity, and informed awareness of the rights, voices, and needs of those who may not be able to be at the table or in the room.

Wonder sparks our imaginations and invites us to care about the Great Lakes. Love forges a commitment to act on their behalf and, in doing so, our own behalf. Grit can manifest in commitment that gives us the courage to act. Wisdom offers guidance on the path.

What wisdom have we gleaned from elders of the Great Lakes? Keep trying. Seek opportunities. Don't close doors. Build relationships. Take the long view. Make friends and have fun. Understand the process by which change is made. Be respectful. Care, learn, and act. Don't take victory for granted; celebrate, and then keep pressing forward. See connections. Know your facts, but let your heart speak its truth. Listen to your intuition. Don't be greedy with nature's gifts and they will continue to sustain you. Acknowledge our dependence on the great diversity of the natural world and express gratitude.

Build community. Seek wisdom, then take responsibility and practice using it. This is global citizenship for the twenty-first century.

CHAPTER TWENTY-TWO

Teach Your Children Well

On this morning, Bill was up early, leaving Colin and me to sprawl across more space in the jumble of sleeping bags and gear inside the tent. There's a certain luxury of two people in a "three-person" tent. Arms and legs can fling out without trespassing into another's allotted space. This, and you can migrate off the inevitable tree root or rock under the tent that you are convinced you alone had to endure last night and seek out a more comfortable spot. But my notions of a languorous morning have been cut short by the percussive (and *loud*) thumping of woodpeckers that are very close by, so I raise my head and scan for them through the mosquito netting.

THUMP-thump-thump-thump-thump. THUMP! They must be so close. Again THUMP-thump-thump-thump-thump—with an accent on the first hit. Suddenly a flash of red and I see not one but two pileated woodpeckers in the ancient birch only thirty yards or so from the tent. Pileated woodpeckers are large birds to begin with, and this close I am struck by their sheer size and bulk.

I'm guessing they are comparable to a crow in size, but strikingly outfitted. The peaked "Woody Woodpecker" crest that our *Peterson Guide* describes as "flaming red" and a dashing white stripe from bill to shoulder on its black body would easily qualify the pileated woodpecker for a fashion "do" in the glamour world of birds. Seeing one is always a rare delight, seeing two this close is the stuff of lasting memories, and I make the parent's choice to create one for my child and hope it lasts.

I rouse the "drowsy moose" of a child next to me and whisper instructions about what I am looking at and point to the tree. The whisper seems gratuitous, as the woodpeckers drill in for another raucous round on the birch. The marvel of it—using one's head as a hammer and chisel and still being able to think and fly afterward.

The birds are oblivious to my puzzling over their anatomy as hardware and are busily and noisily working the tree. As we watch, we observe that one of the birds is much larger than the other and guess that this is a mom woodpecker teaching her little one how to go for the best goodies in the forest. She is rarely more than a few feet away from junior at any point, and she is paying attention to other sounds in the forest in a way that the little one is not. I'm sure she is aware of our presence but has determined we are bird-loving tree huggers, and she isn't paying us too much mind, so we settle in for the show as the sun continues to rise through the trees.

The rising sun is lighting this scene with a golden infusion of warm summer light, and the hint of a breeze is just beginning to stir in the canopy. After many days of rain, the sunlight is startling in its reach through the forest, and we are almost shocked but delighted to see it at last. The stillness of the forest invites listening, and we hear other bird songs and calls in the morning air, but at this moment, the woodpeckers have the area around our campsite to themselves.

Sylvania, like most old-growth forests, provides great woodpecker habitat. Trees that grow large and old and decay on their own time provide both nesting sites and cafeterias to woodpeckers and other wildlife that weave in and out of cycles of the old age and mellow decay of trees. There are gnarly old birches all around us. These aren't the slender white spires of youthful birch but trees that have seen the better part of a century. The bark is yellowed, and the downed wood on the forest floor is usually punky and porous. Various bugs, beetles, and untold legions of microorganisms have mined the best nutritional bits, working happily away at a million meals that will turn a tree into soil over the years. Even so, shreds of the downed bark still work their magic building a campfire, with the stored-up birch oils releasing that extra burst of fuel that can turn a reluctant fire into a little blaze of hope, even on a soggy day.

The woodpeckers are having a great feast on the birches today, but we have seen their distinctive rectangular holes on the big pines and hemlocks throughout our trip as well. Even when we don't see the birds, seeing a woodpecker hole tells us we're in an area where they live and that full cycles of nature are moving in this place. For us, those holes are like a five-star special seal of approval: "This area inspected and approved by pileated woodpeckers." It is a sign of other good things happening in a forest.

There are timber people who have a word for forests like this. They call them "overmature" and speak of waste and loss of board feet of harvestable timber. In the board-foot worldview, an old-growth forest is decay run amok—a lamentable place. But I do not think the birds lament the decay. Where some foresters see waste, the woodpecker sees the stuff of life, nourishment, and rebirth.

As I ponder the old-growth forest, mother bird moves to a higher spot and takes a precise whack at the bark with her sharp and substantial beak, steadying herself, as woodpeckers do, with her sturdy tail feathers. She cocks her head and fixes her eye on junior, as if to say, "Pay attention! Watch this!" She works the area for a while, successfully digging out some unfortunate grub or bark beetle destined for a new role in the carbon cycle. She bounces up and over and watches junior again. The little one comes close to the sweet spot, and she watches him pound away at the tree with apparent success. Having mined some goodies from this spot, she flits to another and again turns her head to make sure of junior's whereabouts and to see whether he is watching.

How do they learn how to be a successful woodpecker? I realize I know nothing about the balance between instinct and learned behavior in woodpeckers, but I am watching and learning now. She certainly seems to be teaching the young one. I imagine she is pointing out the good spots: "Remember, dear, don't waste time on healthy young trees. Go for the old ones with soft spots. Listen for signs of activity under the bark. That's it. Now, brace yourself, pull back your head, and give it a good whack! There you go!"

I look at my own son, who has been watching this display in real time—no fast-forwards, no greatest hits. He knows we are witnessing something special on this golden morning, and the birds have our mutual attention. We sit together, watching from the tent, the stark quiet of the morning magnifying the echo of the pounding birds. I wonder what he is learning on this trip.

On our first canoe trip together, he remarked on the second night: "Out here, I'm totally dependent on you." This was an astute and necessary observation from our precocious six-year-old. We explained that on a trip like this, we were all dependent on each other. Over the years, Bill and I have fallen into an unspoken pattern of making the best use of our different outdoor skills to make our camping adventures more fun and safer, and to get through the chores in efficient ways that maximize "lizard time" on a rock in the sun and minimize fussing over dirty dishes in the dark. We have increasingly woven Colin into the routine, because everyone needs to pitch in for us all to have an enjoyable time. Teamwork and cooperation out here are not something artificially imposed on a ropes course. We truly do depend on each other's knowledge and skill, and we want to pass this on to our next generation.

Colin has watched Bill carefully loading and balancing the gear in the canoe and assessing the wind and water conditions before we set out on open water. Colin observes that we always wear our life jackets, even though we can swim. He has read the fishing rules with Dad and understands that by following the rules, we are protecting the quality of fish and fishing for humans and the loons. He knows that in bear country we always keep a clean camp and hang our food from a line high above our heads, and we respect the lakes and the forests, trying to minimize the impact of our passage through this place.

Here, away from the convenience and comfort of a modern home, I hope he is learning the distinction between the amount of stuff in our daily lives and the simpler essentials of living on what we can carry with us in canoe and pack. He has learned that if your socks get wet, they are going to stay that way for a long time out here. Perhaps he has understood how small three people and a canoe can be in a massive rainstorm, and how we respect (and avoid) lightning in combination with water. He's figured out that whatever we're having for dinner is what's in the pot—even if it was something freeze-dried—and that for safety, pleasure, a warm meal, and a dry bed, we depend on each other.

There are larger lessons for which I hope we are in some way laying the groundwork, about how our species lives wisely on and with the earth. I suppose some of it is instinct, but for humans, a lot of it is learned behavior. If places of clear water and pileated woodpeckers are to remain in our lives, our society needs to learn a lot more about how our lives affect places where the wild edge of nature hangs on by our good graces. Each of us is a teacher of the forest and the lakes, whether we ever set foot on a pine-filled shore or not.

Each of us shares a responsibility to help our next generation learn that we all depend on each other, in the cities as well as the wild places, among and between people and all other living things. A human child needs to know that his or her future is intertwined with the fate of woodpeckers and forests in which they thrive, and ten thousand other miracles of the earth's living fabric every day. This is not a minor fact for an Earth Week quiz; this is a survival skill for the human race. We are totally dependent on the health of a living Earth for every breath we take, all the food we eat, and the quality of the water we drink. Humans, more than any other species on the planet, have damaged and tinkered with the earth's own life-support systems. Why is this so hard to convey? Why do we act like there's a backup planet if we mess things up beyond repair on dear old Earth?

Surely, learning to live lightly and wisely on the planet is in our species' own best interests, as well as of those who share the ride on Earth with us. If we must learn these skills and strategies, we must plumb the deep wells of wisdom across cultures, identify the knowledge, and pass on these skills. It gives new meaning to a different kind of *life* jacket that we need to be aware of, fabricated from knowledge, caring, and the willingness to take responsibility and act. Without it, there's little safety for humankind in the rough waters of the twenty-first century.

So, whether they are lessons about how much stuff we really *need* or where our food comes from, how to care for a river or protect the air we breathe in a city neighborhood, we have much to learn and much to share. Sharing, parent to child, yes, but also sharing in our schools, churches, communities, and institutions. It isn't enough to say that we need to take care of the planet. We must live lives that show how to do it, as individuals and as a society with aspirations of greatness. Children, as the saying goes, learn what they live. My own mother taught me compassion for living things, appreciation for nature's beauty, and not to waste anything. She was the kind of person who would capture a spider or wayward bee in a mason jar and carry it outside rather than kill it. These values were a good foundation for the larger work of my life.

Here, on a gilded morning in a forest by a wild lake, we have been given the gift of watching a mother woodpecker trying her hardest to pass on skills for her young one to not just survive but thrive in this world. As they take flight for their next adventure of the day, it seems to me that, surely, we love our children enough to do as much.

I consider the work that my friend Lee invested in the Dunes Learning Center. She didn't put much value on the old approaches to environmental education that placed so much weight on identifying a leaf or inventorying what can be found in a jar of pond water. The work of the Learning Center builds on experiential learning and the children interacting with the place that is the Indiana Dunes. As Lee put it, "It's about *relationships*." And relationships are fundamental to the way nature works. Why does a certain plant live in a certain place? What kind of soils does it need? What conditions make a favorable habitat? Why does it matter whether a species is native (or not)? How do we fit in?

The current push in U.S. higher education is to develop globally competitive skills in science, math, engineering, and technology. These are all useful, but I hope the teachers in these fields are also teaching about relationships to social issues and the environment, and, frankly, how to think your way out of a wet paper bag when the calculator can't provide the answer to a complicated issue that has moral and ethical implications. Reductionism only gets you so far. It provides a lot of information on "what" and sometimes "how" but not a lot on "now what?" We are missing other key competencies in figuring out how to live on Earth.

I recall a talk by Rita Colwell, former head of the National Science Foundation, who spoke about the need to develop advanced skills in systems thinking to best apply the wealth of knowledge from many fields to the complex system-scale issues before us, whether they are global warming or how to grapple with nanotechnology. Frances Westley, former director of the University of Wisconsin Nelson Institute for Environmental Studies, promoted the idea of "transdisciplinary" approaches. The concept is akin to cross-cultural studies, except it's applied across fields of knowledge.

In a world full of specialization, we need curious and creative thinkers who can knit together the "connective tissue" that enables collective genius to rise from multiple fields along with practical experience and local knowledge that are the feedstock of community wisdom. These "bridging" skills will help us better understand what diverse fields have to offer in the big toolbox of strategies and solutions for environmental problems. Both approaches—systems thinking and transdisciplinary navigation skills—will help foster "relationship literacy." But they don't seem to be getting the wide embrace of leaders in policy or education—at least not yet.

In the elementary schools, the push still seems to be to teach children how to take standardized tests to prevent the loss of precious federal funds—what my own son called the "No Child Left Unstressed" program. We were fortunate that my son's experience included skilled teachers who understood the value of digging into history and exploring the consequences of decision-making as well as the challenge of creating community in a high-speed, fragmented world. These broader areas are the roots of wisdom and survival skills for a new generation faced with devising a strategy to protect the entire fabric of life on Earth.

Research on environmental literacy shows that most Americans are fairly clueless when it comes to how the natural world works and, more importantly, how it relates to them. It is easy to understand why people think water comes from the tap and food comes from the store—that is the evidence of daily experience. If there are raccoons in your garbage and deer in your headlights, people tend to assume that wildlife, and thus nature, must be doing OK. There's the oft-referenced statement about how preschool children can name more commercial brands than types of plants. These are lessons shaped by life in our consumer culture lived largely indoors.

As much as I wish people knew more about the environment, the literacy measures are missing something that the new field of conservation psychology has brought to light. Knowledge in isolation doesn't get you very far unless you *care*. For many years, the Roper Starch annual survey on the environment showed that men scored higher in terms of factual knowledge about topics such as "What is the largest source of oxygen on Earth?" Women knew less about specific facts, but overall they *cared* more about what was happening to the environment and were more likely to take action. So, while factual knowledge is valuable, it isn't the only endpoint in developing a society that can shape solutions and act.

The conservation psychology pioneers are exploring, among other things, what it is about the human psyche that enables us to care for nature—to want to act on behalf of something nonhuman and larger than ourselves. The late biologist E. O. Wilson wrote about biophilia—the love of life—as an innate "hard-wired" part of what makes us human. Author Sara Stein—whose *Noah's Garden* books make the powerful case for restoring ecosystems, starting with our backyards—argued in a lecture that exploring the natural world, even in small ways, is important to human intellectual and emotional development. What toddler (who hasn't already been persuaded that this is dirty or dangerous) doesn't want to touch seed pods in a garden, inspect the caterpillar on the branch, or fill her fists with dirt? Evolutionary biologists might argue that exploring and learning about our environment in the early stages of childhood development is a useful survival skill, but I like to think there is more going on—a forging of relationship and identity. What is self and what is other? How are we connected?

We have a lot of information about what's happening to the Great Lakes and their surrounding environment, but it doesn't necessarily move most people to do anything about this knowledge. The people I know who love the lakes—who understand our connection to them—are motivated to find out what's going on at deeper levels, and they are also the ones likely to show up and be counted in decision-making arenas.

In the challenge to engage more of the people of the Great Lakes region in caring for this place, the lessons seem obvious. Help learners of all ages come to know some portion or experience of the vast region that is their "place." Help them, through experience, understand their relationship with this place and its connection to the wider system. Teach and model caring and celebrate its rewards. Those are beginnings in Great Lakes survival skills that are unlikely to show up as questions on a standardized test. That knowledge, experience, and ability to care might just grow wisdom and leadership—a different measure of the intellectual wealth we, and the precious children who will follow in our footsteps, will need to restore, protect, and live well with the lakes.

PART SEVEN

Reflections

And if dreams could come true, I'd still be there with you
On the banks of cold water at the close of the day.

—From "Keweenaw Light," Craig Johnson

CHAPTER TWENTY-THREE

Loss, Hope, and Courage

My career has been an amazing journey into and through the deep woods of environmental politics and a chance to be part of achieving dreams like the Michigan Wilderness Act and landmark progress to protect the Great Lakes. My conviction rose from a dream for a better world and a higher humanity, a willingness to imagine that which was never before possible in spite of the evidence, and a love for nature and respect for her grand lessons. That's what I've tried to bring with me in the democratic process.

To take these convictions out of the realm of well-intentioned sentiment and put them to work in real social change takes ordinary courage and a willingness to roll up your sleeves and dig into the gritty workings of democracy. For me, that meant learning how to lobby in the halls of Michigan's state legislature and then Congress. It meant going to mind-numbing hearings on technical issues and understanding that process and power relationships were usually more important than facts but that you still had to have your facts down. It meant phone-banking and door-knocking, list building, developing position papers and issue reports, going to political fundraisers, "grip and grin" photo ops with various influentials and decision-makers, countless newsletter articles and sorting envelope labels by zip code. It meant meetings, workshops, opinion research, slideshows, strategy sessions, press releases, letters to the editor, long days and nights away from home, countless bitter compromises, and the occasional victory for the people, the Earth, and future generations.

So, why bother? Why toil along in the slow grind of policy? Because that's what it takes to protect the life-giving systems on earth and their natural beauty and gifts. I am grateful that along my journey, there were moments like the Wilderness tours and site visits, when our work brought us to extraordinary places in the forests, along the shores, and in the vast expanses of the Great Lakes, away from the phones, meetings, and pounding the halls of the House and Senate office buildings in Washington. Those were a rare and treasured perk of the job—times when you could remind yourself what it was all about and recharge your spirit. A favorite line from Bob Franke's song "Thanksgiving Eve" is: "Let your dreams bind your work to your play." By some stroke of grace, I was given that gift, and I hope I used it well.

In the quest to protect the environment, there are losses along the way that we need to process and surmount to keep moving forward. One of mine was Pine Lake. In the 1970s, despite the downturns in Flint's economy, more families built more homes on the lake and began to fill in the remaining lots on the outer rings of the roads that rimmed two-thirds of the lake. Increasingly, people were commuting distances up to sixty miles for the privilege of lake living, and Detroit and Pontiac commuters replaced Flint commuters. Eventually, the county paved the dirt road along the eastern shore, which made even the lots we thought unbuildable attractive to prospective buyers. The developers drained a small wetland and built on other hillsides. Our little rural neighborhood became suburban, and the fields and fallow pastures we had roamed as kids were filling with houses, fences, and more families.

For a long time, the development chipped away, one house at a time. Then, the big blow came in the 1990s, when a developer purchased the western shore that filled our view, including the island. They plunked down ninety houses in a development ironically named Pine Lake Forest on the wooded slope behind the marsh. Next, bulldozers shoved dirt over a culvert and created a road to the island, and they roared onto the island and flattened out building sites where we had run as children. They squeezed six houses on the narrow island on the new Scenic View Drive. The heron's haunt and the lily pad beds became sites for docks, boats, lawn kitsch, and plastic ducks that whirled in gardens to bother the moles.

My dad and I stood on the top of the stairs to the lake the last summer he was there, and we listened to the din of construction noise across the water. He was not a man given to poetic speeches. He simply shook his head and said, "Well, Janie, our paradise is gone." To add salt to the wounds, the developers next swept down upon the undeveloped portions of the other bay on the old camp property. As children, we sometimes bravely ventured back past the bog and into a hidden stand of mature white pine that had somehow escaped the logging boom and must have been part of the reason this was named Pine Lake. The first thing the developers did was log out the white pine.

Soon the bluff above the beach was McMansioned and cul-de-"sacked" and another thirty houses crowded into the little lake's watershed. I know that we were the first wave of sprawl; it was unintentional. My mother just wanted to raise her children in the country. So do thousands of other mothers and fathers, and there is only so much lake country to go around. I don't wish these new families ill on the shores of a different Pine Lake, but I wish they could have known the lake as I knew her, could have run the hills and fields as I did, could have known the dark mysteries of the white pine grove, discovered frogs and turtles in the lily beds, known the clock of the heron on its daily rounds, welcomed home the geese and the blackbirds each spring. Theirs will be a different childhood, more paved, less wild. That is what is there now. It *was* a child's paradise, but for us now, paradise lost, only alive in memory and in my abiding love for lakes. I wasn't in a position to advocate for restraining development on Pine Lake after moving away. Perhaps being a distant witness to those changes that pulled at the heart helped drive my convictions to protect and safeguard other places when the political window was open to do so.

There have been plenty of policy losses along the way, and I have learned from them all. I'll never be able to just shrug them off, but you learn and move forward, sometimes just one step at a time. In the end, I just know that I need to do this work. I'm a different kind of activist today than I was when I was twenty-one, but the roots of who I am today grew from the experiences and impressions of those early years, as they do for so many people. I simply can't passively watch the degradation of this splendid planet and not care.

Along this path, I grew to understand and sometimes question the quirky, slow, flawed, and painstaking processes that our democracy uses to achieve these ends. Some days I think I'm a pragmatic idealist. Other days I see myself as just a laborer in the field who will never see the harvest but still believes there will be one. Then there are days when all the evidence tells you that this is a quixotic quest, and you might as well load up on chocolate and hide under a blanket.

When it all seems utterly hopeless and futile, it's good to withdraw for a while, take care of yourself, and do something that seems real and tangible. Guilt is not allowed, and this is not self-indulgence. Think of it as restoration ecology on a personal scale. In those times, my friend Lee would say, "It's time to make another quilt." And whether it's a quilt, garden, or some other restorative act that affirms life, beauty, and progress, when it is done, there's a thing of beauty. There's also a renewed Lee, Jane, or whoever, ready for the next onslaught—and there will be one.

People ask me whether I'm optimistic, and I say instead that I choose to be hopeful. I have pondered poet Václav Havel's distinctions between the two—optimism being confidence that things will turn out well, and hope being "an ability to work for something because it is good, not just because it stands a chance to succeed."[1] He also says hope is "an orientation of the spirit, an orientation of the heart," and that "it is also this hope, above all, which gives us the strength to live and continually to try new things, even in conditions that seem as hopeless as ours do, here and now."[2]

In recent years, I've heard thoughtful leaders talk about hope being helpful, but courage being essential. During the first two years of the pandemic, I had a little sticker with the word "courage" on it that I had saved from a sheet of Rosa Parks Forever Stamps. I put it on top of my computer monitor so I could see it every morning when I booted up. It helped me stay focused on showing up and getting the essential things done in the swirling uncertainty and anxiety of those times. The *Oxford Dictionary* says courage is the ability to do something that frightens oneself, or strength in the face of pain or grief. Facing climate change and its twin, the extinction crisis, along with all our other environmental challenges, can indeed be frightening—the stakes are high, and the opposition is well funded. These overarching threats can also trigger the pain and grief of known or impending loss. But courage can kindle the strength to keep working anyway, to build alliances, to lift each other up; it is good to strive for life, and beauty, and wholeness, and we don't know where and when the unexpected breakthroughs will come. If we don't try, they most certainly won't. And maybe, just maybe, they will surprise us and reward us and future generations in ways beyond our imagination.

CHAPTER TWENTY-FOUR

Lake Water Lapping

It is the last night of our 2005 trip to Lake Superior. The weather has been splendid and warm, and the lake is calm and lazy, with big, slow waves loping in on a windless shore. Even though the day was full, and we've been packing up and cleaning out the cottage, the lake beckons, and we set aside the chores of the day—there is always tomorrow morning.

The grandchildren visiting at the home next door are riding the swells in the rare warm waters of August, and their laughter is the only encouragement our two ten-year-olds need to plunge in at the sandbar down the beach. The waves are breaking where they run up against the submerged edge of the sandbar, and it makes for a perfect place for the boys to ride and tumble. The shouts and laughter of gleeful children and the rhythm of water rushing in and then subsiding on the shore are like music.

The sun is setting in a cloudless sky, and it looks like it will be a good show over the dark silhouette of Shelter Bay to the west. I grab the sketchbook and observe, taking notes for what might become a painting on a winter's night at home. The silver blue water glides over the pale sands, and sparks of gold reflect in the receding wash as the sun's rays grow long and horizontal and catch the minerals in the sand. I watch the play of light and water and know I'm not a good enough painter to even think about capturing this with pigment and paper. The sky begins that incredible progression, starting with blue, blushes of gold and backlit aqua, then apricot and pink, while the blue deepens in the wider sky. Then, as the sun sinks behind the black trees, flaming *Gone with the Wind* reds and oranges fill the horizon and then slowly fade to burnt oranges, mauves, and the deep purple of late twilight.

The boys are still playing in the lake while the warm air lingers, but as the sky grows darker, wisdom dictates that we should call them out, as any needed rescue in the big lake in the dark seems impossible. And so, we urge them to come in from the embrace of the loving waves, and then finally demand it as the light is fading into the night. Colin is furious with me. He declares: "I am *so* angry. I was having the *best* time of my life and you made me stop!" I offer an apology and a justification, but I mostly hope he remembers this as one of the best times in his life, and that it will help him love this lake forever.

We offer consolation in the form of a fire on the beach, marshmallows, and staying up late enough to watch for meteors. Oh, lovely August, complete with the Perseid meteor showers. Every year, our planet swirls through the cosmic dust of the broad tail of comet Swift-Tuttle, whose backdrop for the light show is the constellation Perseus. In early August, the stardust hits our atmosphere, which just happens to be traveling through the cosmos at sixty-five thousand miles per hour. So, when we make contact with even tiny fragments of comet detritus at these speeds, you can expect an amazing light show as the meteors literally "flame out" and disintegrate. On clear nights, usually in the second week of August, we can see this annual encounter with Swift-Tuttle's tail in one of the best displays of meteors of the year. We hope tonight measures up to the anticipation.

We coax the boys out of sandy swimsuits and into warmer clothes. Bill builds a fire in the ring of stones near the shore that others have clearly used many times before us. We spread out the beach mats and a folding chair or two and are glad for our warmer layers. Bill sharpens sticks with his Swiss Army knife, and the odd blobs of spongy sugar (how did marshmallows become part of campfire rituals?) emerge from their plastic bag and the boys enthusiastically skewer them on their spear-like sticks. The boys have moved from the joy of water to the thrill of fire, turning sugar into flaming carbon torches in the encroaching darkness. Oh, to be a young boy in summer on the shores of a big lake.

We enjoy the waiting, watching night come. First, the bright planets punctuate the sky in a few places, and then, soon, the first-magnitude stars of the summer triangle emerge. Vega shines, bright and blue, straight up overhead in the summer sky. Then we locate the other points of the triangle, Deneb and Altair.

We pick out the easy constellations and stars—the Big Dipper and the North Star, and we "arc" across the handle of the Big Dipper to find the bright star Arcturus at the end. Cassiopeia's chair remains tipped in the low horizon and is easy to find. Soon Cygnus the Swan, also called the Summer Cross, is flying across the Milky Way, with the star Deneb as the tail feather.

On our backs, scanning the sky, our eyes adjust to the starlight. We're hoping that the wisps of clouds starting to form will hold off just a while longer. Then, "*Wow!*" from a young voice as a blazing streak of stardust arcs across the sky and the show begins. Our conversation becomes the shorthand of stargazing. "There's one!" "Over there!" as arms reach up and point to the brief blaze in a corner of the sky. "Did you see *that*?" "Whoa . . . amazing!" Earth is traveling through space, dancing through the meteors that sparkle into our living atmosphere, and we're along for the ride with a front-row seat—two generations loving the moment and witnessing the vivid, fleeting magic of summer. I am having one of my own best times of my life.

Eventually the clouds creep in, and necks tire of craning at odd angles in hopes of catching another streak of light. The beach is damp, and the air takes on a chill. The clouds begin to capture more of the sky, and we know that tomorrow we must leave this special place. We wistfully gather up clammy towels and tired and sticky children, put the fire to bed, and climb up the stairs to the cottage as the waves whisper a farewell in the darkness. We brush the sand off cold feet and dump a heap of beach chairs and towels on the deck to keep most of the damp and sand outdoors. We go through an abbreviated bedtime swish through the bathrooms, and we all slide into various beds, tired and happy. The windows are open, and although we can hear the occasional whine of a car on the highway, it is very quiet.

Up in the night sky, meteors are still raining through the atmosphere, and the stars are moving across their timeless arc of the night. All seems right with the world. I suspect each of us is savoring the thoughts of a day we will long remember on the beautiful shores of the world's greatest lake. I wish for some way to capture this day, this moment, and take it home with me: the sunlight fading into starlight, the unbridled joy of children in the water, the exquisite beauty of sand and stone and clear water on this crescent shore, the fellowship of a campfire, the smell of the damp air that follows twilight, the way sand feels on bare feet. But perhaps the best way to do that is to be fully present in this moment and accept this gift of time and place and magic. In the dark stillness, as we rest our heads on our pillows, we can hear it still: lake water lapping.

Notes

PREFACE

1. William Butler Yeats, "The Lake Isle of Innisfree," Poets.org, https://poets.org/poem/lake-isle-innisfree.

PROLOGUE

1. Government of Canada and U.S. Environmental Protection Agency, *The Great Lakes: An Environmental Atlas and Resource Book*, 3rd ed. (Chicago: U.S. Environmental Protection Agency; Toronto: Government of Canada, 1995), 20. James Proffitt, "Great Lakes Commercial Fishing History," *Great Lakes Now*, November 25, 2019, greatlakesnow.org.

1. PINE LAKE ALMANAC

1. “Interstate Highway System—The Myths,” U.S. Department of Transportation Federal Highway Administration, last updated November 11, 2019, https://www.fhwa.dot.gov.
2. History.com editors, “The 1967 Detroit Riots,” History Channel, last updated March 23, 2021, https://www.history.com/topics/1960s/1967-detroit-riots.
3. Helen Hunt Jackson, “October’s Bright Blue Weather,” Poem of the Week, https://potw.org/archive/potw10.html.
4. “History of the Clean Water Act,” U.S. Environmental Protection Agency, last updated July 6, 2022, epa.gov.

4. AH, WILDERNESS

1. “The Trouble with Wilderness,” William Cronon, williamcronon.net.
2. Tim Gibbins, “An American Original: Aldo Leopold in the Gila Wilderness,” *Your National Forests* (Summer/Fall 2014).
3. Aldo Leopold, “Thinking Like a Mountain,” in *A Sand County Almanac: And Sketches Here and There* (New York: Oxford University Press, 1968), 130.
4. “Our Founders,” The Wilderness Society, wilderness.org.
5. Don Waller, personal communication, September 11, 2022.
6. Max Greenberg, “A Tribute to Howard Zahniser, Unsung Architect of the Wilderness Act,” *The Wilderness Society Blog*, February 25, 2016, wilderness.org.
7. “Other Federal Wilderness Lands,” National Park Service, last updated March 24, 2020, nps.gov.
8. “The Wilderness Act,” Public Law 88-577 (16 U.S.C. 1131-136), 1964.
9. Douglas W. Scott, “‘Untrammeled,’ ‘Wilderness Character,’ and the Challenges of Wilderness Preservation,” *Wild Earth* (Fall/Winter 2001–2002): 72–79.
10. “IUCN Green List of Protected and Conserved Areas,” International Union for the Conservation of Nature, iucn.org.
11. “IUCN Green List of Protected and Conserved Areas.”

12. E. Dinerstein et al., "A Global Deal for Nature: Guiding Principles, Milestones, and Targets," *Science Advances* 5, no. 4 (2019), https://www.science.org/doi/10.1126/sciadv.aaw2869.

5. WELCOME TO THE BIG LEAGUES

1. "Most Remote Spots in USA Wilderness Complexes," Peakbagger.com.
2. "United States—Road Density (Km of Road Per Sq Km of Land Area)," Trading Economics, tradingeconomics.com.
3. Anne Woiwode, "Ah Wilderness! The Halls of Congress, Post 4: Michigan Wilderness Heritage Act, Part 3," *A Quarter Century with the Sierra Club Michigan Chapter*, January 22, 2010, annewoiwode.wordpress.com.
4. Woiwode, "Ah Wilderness!"
5. Woiwode, "Ah Wilderness!"
6. Jane Elder, "The Michigan Wilderness Story: A Personal History," *The Mackinac* 21, no. 1 (Feb./Mar./Apr. 1988): 1, 5–7.

6. WHO OWNS THE SACRED AND BEAUTIFUL?

1. Steve Spicer, "Early Gun Clubs of Northwest Indiana," Steve Spicer's Website, last updated October 2021, spicerweb.org.
2. "Early Development of the Indiana Dunes: 1870s–1910s," National Park Service, last updated December 29, 2022, nps.gov.
3. Lee Botts, discussion with the author, August 1, 2006.
4. Lee Botts, discussion with the author, August 1, 2006.
5. Beth Botts, "Once a Reserve for a Family, but Now a Treasure for All," *Chicago Tribune*, August 28, 2005.

7. PRESERVATION, ENJOYMENT, AND MANAGING EXPECTATIONS

1. "About the National Park Service," National Park Service, https://www.nps.gov/aboutus/aboutus.htm.

2. Theodore J. Karamanski, "Introduction," in *The Pictured Rocks: An Administrative History of Pictured Rocks National Lakeshore* (Omaha: Midwest Regional Office, U.S. Department of the Interior, National Park Service, Midwest Regional Office, 1995).
3. "A Record Year: Pictured Rocks National Lakeshore Reports over a Million Visitors in 2020," *Mining Journal* (Marquette, MI), March 5, 2021, miningjournal.net.
4. Kyle Paoletta, "Give People What They Clearly Need: More National Parks," *New York Times*, August 8, 2021, nytimes.com.

8. NUKES, DUNES, AND THE LADY OF THE LAKES

1. John W. Parsons, *History of Salmon in the Great Lakes, 1850–1970* (Washington, D.C.: U.S. Bureau of Sport Fisheries and Wildlife, Department of the Interior, 1973).
2. *Transfer Federal Water Pollution Control Administration to DOI Is Effected* (Washington, D.C.: U.S. Department of the Interior, May 10, 1966).
3. Lee Botts, discussion with the author, August 1, 2006.
4. Mercer H. Patriarche, *Reaction of Nearby Lake Michigan Fish Stocks to a Heated Discharge at the Palisades Site*, Research Report No. 1820 (Michigan Department of Natural Resources Fisheries Division, Fisheries April 8, 1975).
5. "How Old Are U.S. Nuclear Power Plants, and When Was the Newest One Built?," U.S. Energy Information Administration, last updated March 7, 2022, eia.gov.
6. "Most U.S. Nuclear Power Plants Were Built between 1970 and 1990," U.S. Energy Information Administration, *Today in Energy*, April 27, 2017, eia.gov.
7. Matt Kempner, "How Georgia Nuclear Project's Big Finish Went So Wrong," *Atlanta Journal-Constitution* (Atlanta, GA), December 31, 2021.

9. THE SECOND WAVE

1. Lee Botts and Paul Muldoon, *Evolution of the Great Lakes Water Quality Agreement* (East Lansing, Michigan State University Press, 2005), 28.
2. Gerald Markowitz, "From Industrial Toxins to Worldwide Pollutants: A Brief History of Polychlorinated Biphenyls," *Public Health Reports* 133, no. 6 (2018): 721–25.
3. Botts and Muldoon, *Evolution*, 53.
4. John Jackson, personal communication, July 11, 2022.

10. CHEMISTRY, BIOLOGY, URGENCY

1. Jane Goodall, "Tales from the Food Crisis: The True Horrors of GMOS," *Jane Goodall's Good for All News*, Jane Goodall Institute, October 31, 2016, janegoodall.org.
2. Ronald A. Hites, "Dioxins, an Overview and History, American Chemical Society," *Environmental Science & Technology* 45, no. 1 (September 3, 2010): 16–20. "Dioxins and Their Effects on Human Health," World Health Organization, February 27, 2003, who.int.
3. Riley Beggin, "Graphic Photos Vivid Reminder of Horrors of Michigan PBB Chemical Crisis," *Bridge Michigan*, November 18, 2019, bridgemi.com.
4. Gerald Markowitz, "From Industrial Toxins to Worldwide Pollutants: A Brief History of Polychlorinated Biphenyls," *Public Health Reports* 133, no. 6 (2018): 721–25.

12. TWO STEPS FORWARD

1. "Final Revised Supplemental Finding and Results of the Residual Risk and Technology Review," U.S. Environmental Protection Agency, May 22, 2020, epa.gov.
2. "Proposed Revocation of the 2020 Reconsideration and Affirmation of the Appropriate and Necessary Supplemental Finding," U.S. Environmental Protection Agency, last updated June 8, 2022, epa.gov.
3. "Mercury and Air Toxics Standards (MATS)," U.S. Environmental

Protection Agency, last updated January 31, 2023, epa.gov.

4. *Global Mercury Assessment 2018* (Geneva: UN Environment Programme, Chemicals and Health Branch, 2019), unep.org.
5. Mareike Kroll, *Chemical Hazards in the Artisanal Gold Sector: Impacts of Mercury, Cyanide and Silica Dust on Human Health and the Environment* (Victoria, BC: Artisanal Gold Council, 2020), artisanalgold.org, updated 2022.
6. Keith Ripley, Delia Paul, and Nadia Sohier Zaman, "COP-4.2 Highlights," *Earth Negotiations, Bulletin* 28, no. 61 (March 2022).
7. Danielle Keating, "Study Finds around Half of Great Lakes Residents Know about Advisories Outlining Safe Fish Consumption," Wisconsin Public Radio, July 7, 2022, wpr.org.

13. THE IMAGINED NORTH

1. Edward Abbey, "Joy, Shipmates, Joy!," quoted in *Saving Nature's Legacy: Protecting and Restoring Biodiversity*, by Reed F. Noss, Allen Y. Cooperrider, and Rodger Schlickeisen (Washington, D.C.: Island Press, 1994), 338.
2. Glenda Daniel and Jerry Sullivan, *A Sierra Club Naturalist's Guide to the North Woods of Michigan, Wisconsin, and Minnesota* (San Francisco: Sierra Club Books, 1981), 222.
3. "The New Gate at Crooked Lake," Friends of Sylvania, friendsofsylvania.org.

14. REFUGIA

1. George B. Rabb, "Sounding Off," *EcoHealth* 3 (2006): 70–72.

16. STILL TOXIC AFTER ALL THESE YEARS

1. "Chemicals of Emerging Concern," Human and Ecological Risk Office, California Department of Toxic Substances Control, https://dtsc.ca.gov/emerging-chemicals-of-concern.

2. Greta Fine, Joseph L. Jacobson, Sandra W. Jacobson, Pamela M. Schwarts, and Jeffrey K. Dowler, "Prenatal Exposure to Polychlorinated Biphenyls: Effects on Birth Size and Gestational Age," *Journal of Pediatrics* 105, no. 2 (August 1984): 315–20.
3. J. L. Jacobson, S. W. Jacobson, R. J. Padgett, G. A. Brumitt, and R. L. Billings, "Effects of Prenatal PCB Exposure on Cognitive Processing Efficiency and Sustained Attention," *Developmental Psychology* 28, no. 2 (1992): 297–306.
4. Paul W. Stewart, Edward Lonky, Jacqueline Reihman, James Pagano, Brooks B. Gump, and Thomas Darvill, "The Relationship between Prenatal PCB Exposure and Intelligence (IQ) in 9-Year-Old Children," *Environmental Health Perspectives* 116, no. 10 (October 2008): 1416–22.
5. "Great Lakes Open Lakes Trend Monitoring Program: Polychlorinated Biphenyls (PCBs)," U.S. Environmental Protection Agency, last updated January 5, 2023, epa.gov.
6. M. Harada, "Minamata Disease: Methylmercury Poisoning in Japan Caused by Environmental Pollution," *Crit Reviews in Toxicology* 5, no. 1 (1995): 1–24.
7. "All Wisconsin Waters Have Consumption Recommendations," Wisconsin Department of Natural Resources, dnr.wisconsin.gov.
8. "Canada and the United States Designate the First Set of Chemicals of Mutual Concern," Binational.net, May 31, 2016.
9. Kevin Bunch, "What Are Chemicals of Mutual Concern?," *Great Lakes Connection*, November 9, 2016, ijc.org.
10. "Binational Summary Report: Polychlorinated Biphenyls (PCBs), Draft Summary Draft Document of the Identification Task Team," Binational.net.
11. International Joint Commission, *Atmospheric Deposition of Mercury in the Great Lakes Basin*, 2015.

17. WHITHER THE FISH

1. "Invasive Mussels Now Control a Key Nutrient in the American Great Lakes," University of Minnesota, News and Events, *Research Briefs*, January 26, 2021, twin-cities.umn.edu.

2. "Great Lakes Aquatic Invasive Species," The Nature Conservancy, https://www.nature.org/en-us/about-us/where-we-work/priority-landscapes/great-lakes/great-lakes-aquatic-invasive-species-/.
3. Dan Egan, Great Lakes Revival, "A Watershed Moment, Fishing for a Brighter Future," *Milwaukee Journal Sentinel*, September 2, 2021, jsonline.com.
4. Keith Matheny, "Climate Change Transforming Where Fish in the Great Lakes Region Live," *Detroit Free Press*, September 16, 2019, updated February 11, 2020, freep.com.
5. Sage Marshall, "Wild Lake Trout Fry Found in Lake Erie for First Time in Over 60 Years," *Field & Stream*, August 20, 2021, fieldandstream.com.
6. Morgan Greene, "Lake Superior Is among the Fastest-Warming Lakes on the Planet. Climate Change May Be the Culprit," Phys.org, October 22, 2021.

18. WHEN YOU GET CAUGHT BETWEEN THE MOON AND DETROIT

1. *Greenhouse Gas Emissions from a Typical Passenger Vehicle* (Washington, D.C.: U.S. Environmental Protection Agency, March 2018), epa.gov, last updated June 30, 2022.

19. THEN AND NOW

1. Mimi Abramovitz, "Government Is the Solution, Not the Problem," *New York Times*, March 21, 2021.
2. Richard M. Langworth, "Churchill's 'Democracy Is the Worst Form of Government . . . ,'" Hillsdale College Churchill Project, June 20, 2022, richardlangworth.com.
3. Millennium Ecosystem Assessment, *Ecosystems and Human Well-Being: Synthesis* (Washington, D.C.: Island Press, 2005), 11.
4. "About the Funds," Minnesota's Legacy, www.legacy.mn.gov.

20. IMAGINE THRIVING

1. Daniel Christian Wahl, "Sustainability Is Not Enough: We Need Regenerative Cultures," Medium, March 15, 2017, designforsustainability.medium.com.
2. "Sustainability—What Is It? Definition, Principles and Examples," YouMatter, last modified June 18, 2021, YouMatter.world.
3. Robin Wall Kimmerer, *Braiding Sweetgrass: Indigenous Wisdom, Scientific Knowledge, and the Teachings of Plants* (Minneapolis: Milkweed Editions, 2013), 9.
4. Tim Peterson, "'We Lost a Lot More Than That': The Toll of Losing 40K Wisconsin Dairy Farms in 4 Decades," Wisconsin Public Radio, March 2, 2022, wpr.org.
5. *U.S. Methane Emissions Reduction Action Plan* (Washington, D.C.: White House Office of Domestic Climate Policy, November 2021), whitehouse.gov.
6. John Austin and Alexander Hitch, *A Vital Midwest: The Path to a New Prosperity* (Chicago: Chicago Council on Global Affairs, January 27, 2020).
7. Austin and Hitch, *Vital Midwest*.
8. "Clean Energy Jobs in the Midwest," Clean Jobs Midwest 2021, cleanjobsmidwest.com.

21. WONDER, LOVE, GRIT, AND WISDOM

1. Kathleen Dean Moore, "What It Means to Love a Place," in *The Pine Island Paradox: Making Connections in a Disconnected World* (Minneapolis: Milkweed Editions, 2004), 36.
2. "Thanksgiving Prayer," a version of the traditional Haudenosaunee (Iroquois) thanksgiving address, posted by Allysyn Kiplinger, Ecozoic Times, November 26, 2010, https://ecozoictimes.com/thanksgiving-prayer-from-the-iroquois-nation/.
3. "Thanksgiving Prayer."

23. LOSS, HOPE, AND COURAGE

1. Todd Breyfogle, "Hope by Václav Havel: Reading and Meditation," Aspen Institute, June 2, 2020, aspeninstitute.org.
2. "Disturbing the Peace," Václav Havel Library Foundation, vhlf.org.

Bibliography

Affolter-Caine, Britany, and John C. Austin. *The Vital Center: A Federal-State Compact to Renew the Great Lakes Region*. Washington, D.C.: Brookings Institution, October 1, 2006.

Austin, John, and Alexander Hitch. *A Vital Midwest: The Path to a New Prosperity*. Chicago: Chicago Council on Global Affairs, January 27, 2020.

Botts, Lee, and Paul Muldoon. *Evolution of the Great Lakes Water Quality Agreement*. East Lansing: Michigan State University Press, 2005.

Callicott, J. Baird, and Michael P. Nelson. *The Great New Wilderness Debate*. Athens: University of Georgia Press, 1998.

Colborn, Theo, Dianne Dumanoski, and John Peterson Myers. *Our Stolen Future: Are We Threatening Our Fertility, Intelligence, and Survival? A Scientific Detective Story*. New York: Dutton/Penguin Books, 1996.

Daniel, Glenda, and Jerry Sullivan. *A Sierra Club Naturalist's Guide to the North Woods of Michigan, Wisconsin, and Minnesota*. San Francisco: Sierra Club Books, 1981.

Dinerstein, E., T. E. Lovejoy, J. Mayorga, D. Olso, G. P. Asner, E. M. Baillie, N. D. Burgess, et al. "A Global Deal for Nature: Guiding Principles, Milestones, and Targets." *Science Advances* 5, no. 4 (2019). https://www.science.org/

doi/10.1126/sciadv.aaw2869.

Elder, Jane. "The Michigan Wilderness Story: A Personal History." *The Mackinac* 21, no. 1 (Feb./Mar./Apr. 1988).

Emerick, Lon L. *Going to Back to Central: On the Road in Search of the Past in Michigan's Upper Peninsula*. Skandia: North Country Publishing, 2003.

Fine, Greta, Joseph L. Jacobson, Sandra W. Jacobson, Pamela M. Schwarts, and Jeffrey K. Dowler. "Prenatal Exposure to Polychlorinated Biphenyls: Effects on Birth Size and Gestational Age." *Journal of Pediatrics* 105, no. 2 (August 1984): 315–20. https://doi.org/10.1016/S0022-3476(84)80139-0.

Gibbins, Tim. "An American Original: Aldo Leopold in the Gila Wilderness." *Your National Forests* (Summer/Fall 2014).

Government of Canada and U.S. Environmental Protection Agency. *The Great Lakes: An Environmental Atlas and Resource Book*. 3rd ed. Chicago: U.S. Environmental Protection Agency; Toronto: Government of Canada, 1995.

Harada, M. "Minamata Disease: Methylmercury Poisoning in Japan Caused by Environmental Pollution." *Critical Reviews in Toxicology* 25, no. 1 (1995): 1–24. https://doi.org/10.3109/10408449509089885.

Hites, Ronald A. "Dioxins, an Overview and History." *Environmental Science & Technology* 45, no. 1 (2010): 16–20. https://doi.org/10.1021/es1013664.

Hogue, Cheryl. "Endosulfan Banned Worldwide." *Chemical and Engineering News*. May 5, 2011.

International Joint Commission. *Atmospheric Deposition of Mercury in the Great Lakes Basin*. December 2015.

Jacobson, J. L., S. W. Jacobson, R. J. Padgett, G. A. Brumitt, and R. L. Billings. "Effects of Prenatal PCB Exposure on Cognitive Processing Efficiency and Sustained Attention." *Developmental Psychology* 28, no. 2 (1992): 297–306. https://doi.org/10.1037/0012-1649.28.2.297.

Karamanski, Theodore J. *A Nationalized Lakeshore: The Creation and Administration of Sleeping Bear Dunes National Lakeshore*. U.S. Department of the Interior, National Park Service, 2000.

Karamanski, Theodore J. *The Pictured Rocks: An Administrative History of Pictured Rocks National Lakeshore*. Omaha: U.S. Department of the Interior, National Park Service, Midwest Regional Office 1995.

Kimmerer, Robin Wall. *Braiding Sweetgrass: Indigenous Wisdom, Scientific Knowledge, and the Teachings of Plants*. Minneapolis: Milkweed Editions, 2013.

Lauritsen, Marv, and Norm Jean Lauritsen. *Sylvania: Where Man Is a Visitor*. Rhinelander: Jensmarie Publishing, 2006.

Leopold, Aldo. "Thinking Like a Mountain." In *A Sand County Almanac and Sketches Here and There*, 129-133. New York: Oxford University Press, 1987.

Lopez, Barry. "An Interview with Oren Lyons." *Orion Magazine* (January/February 2007).

Markowitz, Gerald. "From Industrial Toxins to Worldwide Pollutants: A Brief History of Polychlorinated Biphenyls." *Public Health Chronicles* 133, no. 6 (2018). https://journals.sagepub.com/doi/10.1177/0033354918801578.

Millennium Ecosystem Assessment. *Ecosystems and Human Well-Being: Synthesis*. Washington, D.C.: Island Press, 2005.

Moore, Kathleen Dean. *The Pine Island Paradox: Making Connections in a Disconnected World*. Minneapolis: Milkweed Editions, 2004.

National Park Service. *Great Lakes, Remaining Shoreline Opportunities in Minnesota, Wisconsin, Illinois, Indiana, Ohio, Michigan, Pennsylvania, New York, by the Great Lakes Shoreline Recreation Area Survey*. Washington, D.C.: National Park Service, 1959.

Noss, Reed, and Allen Cooperrider. *Saving Nature's Legacy: Protecting and Restoring Biodiversity*. Washington, D.C.: Island Press, 1994.

Parsons, John W. *History of Salmon in the Great Lakes, 1850–1970*. Washington, D.C., Department of Interior, U.S. Bureau of Sport Fisheries and Wildlife, 1973.

Patriarche, Mercer H. *Reaction of Nearby Lake Michigan Fish Stocks to a Heated Discharge at the Palisades Site*. Research Report No. 1820. Michigan Department of Natural Resources Fisheries Division, Fisheries, April 8, 1975.

Peacock, Bonnie, *Sylvania Majestic Forests and Deep, Clear Waters*. Waite Park: Park Press, 1986.

Rabb, George B. "Sounding Off." *EcoHealth* 3 (2006): 70–72.

Ripley, Keith, Delia Paul, and Nadia Sohier Zaman. "COP-4.2 Highlights." *Earth Negotiations, Bulletin* 28, no. 61 (March 2022).

Scott, Douglas W. "'Untrammeled,' 'Wilderness Character' and the Challenges of Wilderness Preservation." *Wild Earth Magazine* (Fall/Winter 2001–2002): 72-79.

Scott, Douglas W. *The Enduring Wilderness*. Wheat Ridge: Fulcrum Press, 2004.

Stewart, Paul W., Edward Lonky, Jacqueline Reihman, James Pagano, Brooks B. Gump, and Thomas Darvill. "The Relationship between Prenatal PCB

Exposure and Intelligence (IQ) in 9-Year-Old Children." *Environmental Health Perspectives* 116, no. 10 (October 2008): 1416–22. https://ehp.niehs.nih.gov/doi/10.1289/ehp.11058.

United Nations Environment Programme. *Global Mercury Assessment 2018*. Geneva: UN Environment Programme, Chemicals and Health Branch, 2019.

United Nations Minamata Convention on Mercury. *Minamata Convention in 2022: Progress Report on Activities*. Geneva: UN Environment Programme, 2023.

U.S. Department of the Interior. *Transfer Federal Water Pollution Control Administration to DOI Is Effected*. Washington, D.C.: U.S. Department of the Interior, May 10, 1966.

Index

C

D